NEBOSH NATIONAL DIPLOMA

UNIT A: MANAGING HEALTH AND SAFETY – PART 2

Element A7: The Assessment and Evaluation of Risk

Element A8: Risk Control

Element A9: Organisational Factors

Element A10: Human Factors

Element A11: The Role of The Health and Safety Practitioner

CONTRIBUTORS

Craig Davies, DipNEBOSH, CMIOSH, MIIRSM, MIET, FETC, PIEMA, TechIFE, MInstL&M, PTTLS, ISO Lead Auditor/Trainer

With thanks to:

Dr Terry Robson, Bsc (Hons), PhD, CFIOSH, MRSC, CChem

Dr David Wenham, BSc, MSc, PhD, DipSH, DipLaw, AdvDipEd, MRSC, CChem, CFIOSH

For information on all RRC publications and training courses, visit: www.rrc.co.uk

RRC: NDA.2 Part 2

ISBN for this volume: 978-1-911002-75-8
Fourth edition May 2021

ACKNOWLEDGMENTS

RRC International would like to thank the National Examination Board in Occupational Safety and Health (NEBOSH) for their co-operation in allowing us to reproduce extracts from their syllabus guides.

This publication contains public sector information published by the Health and Safety Executive and licensed under the Open Government Licence v.3 (www.nationalarchives.gov.uk/doc/open-government-licence/version/3).

Every effort has been made to trace copyright material and obtain permission to reproduce it. If there are any errors or omissions, RRC would welcome notification so that corrections may be incorporated in future reprints or editions of this material.

Whilst the information in this book is believed to be true and accurate at the date of going to press, neither the author nor the publisher can accept any legal responsibility or liability for any errors or omissions that may be made.

Contents

Introduction

Element A7: The Assessment and Evaluation of Risk

Contents

Element A8: Risk Control

Element A9: Organisational Factors

Contents

Element A9: Organisational Factors (Continued)

Element A10: Human Factors

Contents

Element A11: The Role of the Health and Safety Practitioner

Revision and Examination

Suggested Answers

Course Structure

This textbook has been designed to provide the reader with the core knowledge needed to successfully complete the NEBOSH National Diploma in Occupational Health and Safety, as well as providing a useful overview of health and safety management. It follows the structure and content of the NEBOSH syllabus.

The NEBOSH National Diploma consists of four units of study. When you successfully complete any of the units you will receive a Unit Certificate, but to achieve a complete NEBOSH Diploma qualification you need to pass the three units within a five-year period. For more detailed information about how the syllabus is structured, visit the NEBOSH website (www.nebosh.org.uk).

Assessment

Unit A is assessed by a two-part, three-hour exam. Section A consists of six 10-mark compulsory questions, and Section B consists of five 20-mark questions, of which you must choose three.

NEBOSH set and mark this exam paper.

More Information

As you work your way through this book, always remember to relate your own experiences in the workplace to the topics you study. An appreciation of the practical application and significance of health and safety will help you understand the topics.

Keeping Yourself Up-to-Date

The field of health and safety is constantly evolving and, as such, it will be necessary for you to keep up to date with changing legislation and best practice.

RRC International publishes updates to all its course materials via a quarterly e-newsletter (issued in February, May, August and November), which alerts students to key changes in legislation, best practice and other information pertinent to current courses.

Please visit www.rrc.co.uk/news/newsletters.aspx to access these updates.

Unit A: Managing Health and Safety	
Element A1	Principles of Health and Safety Management
Element A2	Principles of Health and Safety Law
Element A3	Criminal Law
Element A4	Civil Law
Element A5	Loss Causation and Incident Investigation
Element A6	Measuring and Reviewing Health and Safety Performance
Element A7	The Assessment and Evaluation of Risk
Element A8	Risk Control
Element A9	Organisational Factors
Element A10	Human Factors
Element A11	The Role of the Health and Safety Practitioner (Please note that Element A11 will **NOT** be examined in the Unit A exam, but only assessed as part of the Unit DNI assessment.)

The Assessment and Evaluation of Risk

Learning Outcomes

Once you've read this element, you'll understand how to:

1. Explain how to use internal and external information sources in identifying hazards and the assessing of risk.

2. Outline the use of a range of hazard identification techniques.

3. Explain how to assess and evaluate risk and to implement a risk assessment programme.

4. Explain the analysis, assessment and improvement of system failures and system reliability with the use of calculations.

5. Explain the principles and techniques of failure tracing methodologies with the use of calculations.

Contents

Sources of Information Used in Identifying Hazards and Assessing Risk

IN THIS SECTION...

- Accident and ill-health data may be used to calculate incidence, frequency, severity and prevalence rates.

- External information sources include:

    - Health and Safety Executive (HSE).

    - European Safety Agency.

    - International Labour Organization (ILO).

    - World Health Organization (WHO).

    - Professional bodies, e.g. Institution of Occupational Safety and Health (IOSH).

    - Trade bodies.

- Internal information sources include:

    - Injury data.

    - Ill-health data.

    - Property damage.

    - Near-miss information.

    - Maintenance records.

- Accident, ill-health and near-miss data are often under-reported.

- Trends cannot be established unless large amounts of data are collected over relatively long periods of time.

Accident/Incident and Ill-Health Data and Rates

Incident data can be used to support hazard identification, risk assessment and risk factors. Some of the accident data collected is used to generate the statistics required by legislation, but other uses are to:

- Classify industries according to risk.

- Classify workplaces.

- Classify occupations.

- Consider accident trends.

- Consider parts of the body injured – use of protective clothing.

- Use 'cause of injury' to determine hazards in a workplace.

- Consider where the fault lies.

- Measure the effect of preventive/control measures.

Incident data can support hazard identification

It can sometimes be difficult to obtain and interpret such information (because of assumptions, etc.).

Accident and Disease Ratios

In making comparisons between various industries, or between work areas in the same factory, it is useful to consider the commonly used accident and disease rates. You will remember these from Element A5.

Incidence Rate

Incidence reflects the number of new cases in relation to the number at risk.

$$\frac{\text{Number of work-related injuries}}{\text{Average number of persons employed}} \times 1{,}000$$

It is a measure of the number of injuries per 1,000 employees and is usually calculated over a period of time, e.g. a year. It is often applied to discrete events, such as accidents.

Frequency Rate

$$\frac{\text{Number of work-related injuries}}{\text{Total number of hours worked}} \times 100{,}000$$

It is a measure of the number of accidents per 100,000 hours worked.

Severity Rate

$$\frac{\text{Total number of days lost}}{\text{Total number of hours worked}} \times 1{,}000$$

It is a measure of the average number of days lost per 1,000 hours worked.

Prevalence Rate

Prevalence is a term often used to describe ill health in terms of the proportion of persons who have the prescribed ill-health condition at a particular time.

$$\frac{\text{Total number of cases of ill health in the population}}{\text{Total number of persons at risk}} \times 100$$

This calculation will give the percentage of the population with the disease.

External Information Sources

TOPIC FOCUS

You need to be aware of the help that is available to the safety practitioner when carrying out risk assessment.

Health and Safety Executive

The major source of accident statistics and other information is the HSE. (In information produced before December 2008 you may also find reference to the Health and Safety Commission (HSC), which has since merged with the HSE.)

- The **Annual Report** is an important publication, but there are also many guidance notes and pamphlets.

- **HSE inspectors** are a great source of advice, and regard this, rather than prosecution, as their main duty.

- Within the HSE Report, there are data from the Department for Work and Pensions on accidents and ill-health numbers which relate to people in employment who are claiming benefit.

- **Death certificates** will often record information suggesting an industrial disease or condition as a factor. This information is in the HSE Report, as well as details of special investigations.

Each year, the HSE publishes a standard format annual report giving details of the previous year's reported injuries and ill health. The figures are always provisional because, where a person dies within one year of the injury, it is counted as a fatality. A separate statistical supplement is also often published.

There are two separate publications: the **HSE Annual Report** and the **HSE Health and Safety Statistics**. The report indicates trends and enables the HSE inspectors to concentrate their efforts on those areas where there is need for greater improvement.

The **Labour Force Survey** is another source of information which is often quoted in the HSE report. The information comes from answers to a questionnaire where the data is based on the respondents' own perceptual link between their ill health and their working situation. Here, a distinction is made between conditions **caused** by work and those which are **made worse** by the work situation.

More men than women report work-related illness in all age groups. For those over retirement age, there are three times more men than women reporting. In the 16 to 44 age group, the figure is only one third more.

Can you think of reasons why this could be so?

There are wide differences between the types of ill health reported by manual workers, farmers, nurses, office workers and, the armed forces and other professions.

For example:

- Stress/depression is the main cause of ill health for teachers.

- Skin disease and deafness are problems in science and engineering.

- Varicose veins are common among saleswomen staff.

It is suggested that 900,000 days are lost every year due to skin troubles or industrial dermatitis.

Separate lists are available for people working at mines, quarries and railways. Although it is unlikely that examination questions will be set on these specialist areas, you need to know about them if you work in these industries.

(Continued)

TOPIC FOCUS

Internal information is the most relevant data for an organisation, but other sources will be needed for comparisons, unless the organisation is large enough to give statistical significance.

There should be a source of accident and ill-health data, as well as near-miss information, within the company:

- **Accident reports** will be the most obvious source. It is important that the information recorded is adequate for risk assessments. We need to be able to investigate factors that contributed to the accident, which means making a clear distinction between '**cause of accident**' and '**cause of injury**'.

- **Absence records** may be another indication of problems. Health problems may not always be reported, so conditions made worse by the work situation, rather than being caused by it, are not so easy to spot.

- **Maintenance records** will usually show **damage incidents**.

Other National Sources of Information

- Statistical information is usually available from industry sources. This varies in its nature and usefulness. In some industries, there is great emphasis on safety and health matters and the statistics will include tables that compare the various factors. This information is useful in identifying those hazards that need to be controlled, and the solutions that have been found to be most effective. Comparisons can be made between industry averages and your own achievements.

Statistical information is available from a variety of sources

- A number of **trade unions** produce information on safety and health matters. The trade union interest here may be in making members aware of possible compensation areas. Even though the safety practitioner is usually a member of the management of the organisation, they may have to advise employees on their rights to compensation.

- **Insurance companies** set the levels of premiums and need data to calculate the probable risks of any venture. For example, while the individual lifespan is most unpredictable, the probability of dying is very certain, and the average lifespan is well known.

The average risks involved in most activities can be found in insurance tables. Since the risk manager is involved in managing risks, these tables will be extremely useful, although getting hold of them may not be so easy.

MORE...

The European Agency for Safety and Health at Work has published *Health and safety risks at the workplace: A joint analysis of three major surveys*, which provides an in-depth picture of how safety and health is managed in Europe's workplaces, the perspectives of establishments on risk management and risk awareness, and those of workers on exposure to risks and occupational safety and health outcomes.

International Sources

Both the **ILO** and the **EU** produce literature which is probably most useful to international organisations, but it gives a worldwide view and indicates the co-operation between states on health and safety matters. A specialised industrial concern may need to use worldwide data to obtain a significant information base. This source would usually be found in a library.

The **WHO** works to provide the needed health and well-being evidence through a variety of data collection platforms, including the World Health Survey covering almost 400,000 respondents from 70 countries, and the *Study on global aging and adult health* (SAGE) covering over 50,000 persons over 50 years old in 23 countries. The Country Health Intelligence Portal (CHIP) has also been developed to provide an access point to information about the health services that are available in different countries. The information gathered in this portal is used by the countries to set priorities for future strategies or plans, implement, monitor, and evaluate it.

A **professional** association (also called a **professional body**, **professional organisation**, or **professional society**) seeks to further a particular profession, the interests of individuals engaged in that profession and the public interest.

For example, IOSH is the world's largest health and safety membership body. With 47,000 members in more than 130 countries, it is committed to ensuring that global work practices are safe, healthy and sustainable.

As with most professional bodies, its resources offer everything from clear, accessible introductions to subjects, to full coverage of health and safety specific issues.

MORE...

If you visit the IOSH website you will find resources on health and safety management, environmental management, corporate social responsibility and more available at:

www.iosh.co.uk

Uses and Limitations of Information Sources

Internal information is obviously very relevant to risk assessments. However, the absence of accidents is not a very good indication that all is well. Can you think why this is so?

The absence of accidents is not necessarily an indication all is well

Accidents should be rare occurrences. Quite often, there is a large element of chance involved in the severity of an accident. Near misses, which are usually a much larger figure, are a better indicator.

Care must be taken when using **external sources**. The numbers are larger, and any statistics are based on a larger sample, so are statistically more relevant. However, the type of industry covered may be much wider than your own situation. In the case of a very specialised situation, this may be the only indication of risk available. Different sources use a different multiplier when working out accident frequency rates, etc. so care needs to be taken when making comparisons.

When comparing data between organisations, it is important to make sure that they have the same terms of reference. For example, when comparing Lost-Time Accident (LTA) incidence rates between two organisations (based on numbers of employees), note the following:

- The two organisations may use different definitions for an LTA (many companies use > one day lost for internal reporting of LTA, rather than the **Reporting of Injuries, Diseases and Dangerous Occurrences Regulations (RIDDOR)** seven-day and three-day definitions).

- There is no indication of injury severity.

- The figures may be for employees only and so may be misleading for an organisation that makes wide use of contractors. The figures may not take full account of overtime or part-time staff (they may not adjust the numbers of employees to 'full-time equivalents').

- Culture differences – one organisation may have a culture where they take time off even after a very minor injury; another organisation might have a very strong 'back to work' culture where an injured worker might be brought back to work on restricted or 'light' duties in order to avoid recording an LTA.

- There may be different risk levels between organisations due, for example, to the nature of the work, premises, equipment, etc.

- There may be different risk management arrangements in place relating, for example, to standards of risk assessment, training requirements and standards of control.

STUDY QUESTIONS

1. What uses might we have for accident and ill-health data?

2. What internal information can organisations use to help in the assessment of risk?

3. Explain the difference between 'incidence' and 'prevalence' when referring to accident or ill-health statistics.

(Suggested Answers are at the end.)

Hazard Identification Techniques

IN THIS SECTION...

Hazards may be identified using:

- Observation.
- Task analysis.
- Checklists.
- Incidents and near misses.
- Failure tracing techniques.

Detection of Hazards

A hazard is anything that may cause harm, such as chemicals, electricity, working from ladders, an open drawer, etc. It is important, at Diploma level, to appreciate the difference between activities and hazards – for example, work at height is an activity in which there may be a number of hazards involved, such as an unprotected edge, a damaged ladder, high winds, etc. So, while a term like 'work off a ladder' may be included in the 'Five Steps to Risk Assessment', it is an oversimplification.

Before risk can be managed effectively, any hazards in the workplace have to be identified. Hazards missed at this stage will not be considered later. There are various techniques that can be used to detect hazards, including: observation of tasks, task analysis, use of checklists during inspections, and incident reports.

Hazards in the workplace must be identified

Observation

MORE...

HSE guidance on risk assessment is contained in INDG163(rev4) *Risk assessment – A brief guide to controlling risks in the workplace*, available at:

www.hse.gov.uk/pubns/indg163.pdf

Many hazard identification techniques rely on observation by the assessor(s) and are dependent on the experience and knowledge of the assessor.

The analyst should observe the work being done, including work being carried out by groups of operators, looking for:

- Actual and potential hazards – by observation and questioning.
- Less obvious, 'invisible' hazards – such as health dangers from fumes, gases, noise, lighting, etc.
- Behavioural aspects – rules and precautions for controlling any hazard or risk should be supplied by the operator, their supervisor or a specialist, but are they being followed?

Task Analysis

Task analysis is used to analyse all aspects of a task (including safety), often with the intention of improving efficiency. A job can also be analysed with the emphasis on safety or hazards.

The assessor divides the task into a number of steps, considering each step separately. The results of this analysis can be used to correct existing problems and to improve, among other things:

- Safe working methods, working instructions, worker protection, safety rules, emergency procedures, serviceability of machinery and plant.

- Reporting of hazards, provision of information.

- Layout of work areas.

Checklists

To ensure a consistent and comprehensive approach to checking all the safety elements to be covered during an inspection, a checklist or inspection form is usually developed which covers the key issues. Checklists should also be structured to provide a coherent approach to the inspection process. This helps in the monitoring of the inspection process and analysis of the results, as well as simplifying the task of carrying out the inspection itself. Checklists do have some limitations in that although they prompt the assessor when looking for hazards, any hazard **not** identified in the list is less likely to be noticed.

The HSE has promoted the following '4 Ps' structure to cover the four areas concerned with work activities and risk creation:

- **Premises**, including:
    - Access/escape.
    - Housekeeping.
    - Working environment.
- **Plant and substances**, including:
    - Machinery guarding.
    - Local exhaust ventilation.
    - Use/storage/separation of materials/chemicals.
- **Procedures**, including:
    - Permits to work.
    - Use of personal protective equipment.
    - Procedures followed.
- **People**, including:
    - Health surveillance.
    - People's behaviour.
    - Appropriate authorised person.

Checklists help in the monitoring and inspection process

(Note that the examples given are purely for illustration and are not intended to be a definitive list.)

While checklists are often included in safety procedures and manuals, do not feel that they cannot be changed and adapted. In particular, in terms of maintenance and safety inspections, the list should not act as a constraint on the inspector(s) identifying other potential problems or hazards. Checklists should be reviewed regularly to take account of recent or proposed developments in health and safety issues in the particular workplace.

Incident Reports

These represent reactive, but nonetheless useful, data. Reports can arise from external reporting requirements under **RIDDOR** (see Element A5) or from internal reporting of all accidents (not just **RIDDOR** events). Certain accidents and occupational ill-health incidents are reported to, and analysed by, the Health and Safety Inspectorate to find out the circumstances and total fatalities in industrial processes. Annual reports are published giving detailed figures of **reported** accidents 'analysed by cause'. These are useful when comparing differing sectors of industry.

Each company should maintain its own records, not just of 'reportable' accidents and ill health, but of all accidents that have taken place. In this way, any trends or particular areas that show significant changes can be investigated at the earliest opportunity. The reports will also be a useful tool when carrying out risk assessments.

Failure Tracing Techniques

The techniques we have described are usually more than adequate for most risk assessments. In more complex systems, it may be necessary to use more structured methods to identify hazards. One powerful technique is called a Hazard and Operability Study (HAZOP).

This breaks down a system such as a chemical process into different sections and then systematically asks what could go wrong in that section, what would be the consequences, and what measures could be introduced to reduce the likelihood of the failure occurring or, if it does fail, might mitigate the consequences. We will look at this and other techniques later in this element.

MORE...

You can find more information on the extensive range of hazard identification techniques in the Health and Safety Laboratory report: *HSL/2005/58 – Review of hazard identification techniques*

This is accessible at:

www.hse.gov.uk/research/hsl_pdf/2005/hsl0558.pdf

Importance of Employee Input

Rather than relying on one individual to undertake hazard identification, the HSE advocate the team approach to risk assessment. It is important to involve employees who have relevant experience and knowledge of the process or activity being considered as they are likely to have the best understanding of the hazards.

Can you think of any other reasons why employees should be involved?

Involving employees also increases the 'ownership' of the assessment as, having contributed to the exercise, an individual is more likely to appreciate the need for compliance with the control measures identified.

STUDY QUESTION

4. Giving two examples for each, identify the '4 Ps' recommended by the HSE when preparing a checklist for inspections.

(Suggested Answer is at the end.)

Assessment and Evaluation of Risk

IN THIS SECTION...

- The key steps in a risk assessment are:
  - Hazard identification.
  - Identification of who is at risk.
  - Estimation, evaluation of risk and identification of precautions.
  - Recording of significant findings and implementation.
  - Reviewing the assessment.
- Risk assessments may be generic, specific or dynamic.
- Risk assessments should take account of temporary and non-routine operations and consider long-term health hazards.
- Risk assessments may be qualitative, semi-quantitative or quantitative.
- The risk assessment process should be reflected in a policy, organising, planning and implementing, monitoring and review.

Key Steps in a Risk Assessment

According to the HSE publication *Risk assessment – A brief guide to controlling risks in the workplace* (INDG163):

> "A risk assessment is not about creating huge amounts of paperwork, but rather about identifying sensible measures to control the risks in your workplace."
>
> Source: HSE (www.hse.gov.uk/risk/controlling-risks.htm)

For the majority of work activities, the HSE recommends an approach based on the following five steps:

1. Identify the hazards.
2. Decide who might be harmed and how.
3. Evaluate the risks and decide on precautions.
4. Record your significant findings and implement them.
5. Regularly review your assessment and update if necessary.

Implicit within Step 3 of risk evaluation is the need to estimate the magnitude of the risk so here we will consider Step 3 as risk estimation, evaluation and deciding on precautions.

Step 1 — Identify the hazards

Step 2 — Identify the people who might be harmed and how

Step 3 — Evaluate the risk and decide on precautions

Step 4 — Record the significant findings

Step 5 — Review and update as necessary

The UK's HSE's steps

Comprehensive Identification of Risks

If risk assessment is about identifying sensible measures to control the risks in the workplace, then the starting point has to be a comprehensive identification of risks. The range of risks in an organisation is a key factor in determining the approach that needs to be taken to manage health and safety, and the 'riskier' the organisation, the more effort is needed to manage those risks. Each organisation has specific risks that arise from the nature of the business. In some businesses, the risks will be tangible and arise from safety issues, whereas in other organisations, the risks may be health-related and longer-term.

A broad examination of the nature and level of the threats faced by the organisation and the likelihood of these adverse effects occurring (i.e. severity and likelihood) will establish the likely level of disruption and cost associated with each type of risk. This enables significant risks to be identified and prioritised for action with minor risks simply noted to be kept under review.

Identify the Hazards

This is a crucial step in any risk assessment; we have already discussed the techniques that can be used.

Identifying Persons at Risk

It is important to identify the different categories of persons who are exposed to each hazard, because this will influence the choice of control measures that could be adopted to reduce the risk. For example, control measures such as training could be used to protect employees, but would not be practicable for protecting non-employees, such as members of the public.

The categories might include:

- Employees carrying out a task, e.g. operating a lathe.
- Other employees working nearby who might be affected.
- Visitors/members of the public.
- Maintenance staff.
- New/young workers.
- Persons with a disability.
- Persons who work for another employer in a shared workplace.

For each category, you need to identify how they might be harmed. For example, for employees operating a lathe, loose clothing could become entangled in the rotating spindle, and other employees working nearby might be struck by swarf.

Risk Estimation, Evaluation and Precautions

DEFINITIONS

RISK ESTIMATION

Risk estimation is determining the magnitude of the size of the risk.

RISK EVALUATION

Risk evaluation is the process of deciding whether a risk is acceptable or not.

When we refer to 'risk' in relation to occupational safety and health, the most commonly used definition is: 'the likelihood of a person being harmed or suffering from any adverse health effects when exposed to a hazard.'

It is important that you understand the difference between **risk estimation** and **evaluation**.

Risk estimation may range from being a relatively crude estimation, e.g. high, medium or low, to a more accurate estimation based on data. 'Estimation' is used because risk deals with uncertainty and even the most detailed risk assessments have to make a number of assumptions.

Factors Affecting Probability and Severity of Risk

The magnitude of a risk associated with an incident is determined by two factors:

- the likelihood or probability of the event occurring, and

- the consequence or harm realised if the event takes place. This is usually expressed as:

$$\text{Risk} = \text{Likelihood (or Probability)} \times \text{Consequence (or Harm)}$$

We now need to look at the factors that affect both the likelihood and the consequence.

The **likelihood** of an adverse event occurring is affected by two factors:

- degree of exposure to the hazard and, once exposed to the hazard, and

- the likelihood that harm will occur.

Let's look at an example.

Consider a torn carpet in an office and the risks it creates. Before somebody could possibly trip on the carpet, they have to walk in the vicinity of the carpet, so the degree of exposure to the hazard is a key factor. If the carpet is situated in the centre of a main walkway then the likelihood of it causing an accident is much greater than if it is in a corner of a little-used store room. Similarly, the hazard from crossing a road will create a greater probability of harm if we cross the road several times a day rather than if we cross it only several times a year.

Of course, merely encountering a hazard does not mean we will be harmed. Some people will see the hazard and avoid it deliberately, and others will walk over it without being tripped up. In other words, a specific set of events must occur before a trip will occur, namely:

- Not noticing the hazard and taking avoiding action.

- Placing a foot into the tear such that the person can no longer maintain their balance.

The **consequence** is the outcome from the adverse incident occurrence. From most such incidents there is not just one possible outcome, but a whole range of them. For example, with our torn carpet, one individual might trip on the tear, recover their balance and suffer no harm at all. At the other extreme, someone else may trip, hit their head and die. In practice, we have to use our judgment to decide the most likely outcome, probably in this case just a bruise. However, if the same incident occurred in a care home where most of the residents were elderly and frail, then the most likely consequence may well be a fracture.

Risk Evaluation

Having estimated the magnitude of the risk, we then have to decide if the existing control measures are adequate or whether additional/different ones are necessary. According to the HSE publication *Risk assessment – A brief guide to controlling risks in the workplace* (INDG163), the easiest way of evaluating the risk is to compare your practices with recognised guidance. This is more than adequate for most assessments. However, as health and safety practitioners who may encounter non-routine or more complex risks, having made an informed estimation of the magnitude of the risk, we can use the information to help us decide what action, if any, is necessary.

Risk Control Standards

Having evaluated the risk and established whether or not it is acceptable, you have to ensure that the controls meet minimum standards. Such standards may be defined in:

- Legislation.

- Approved Codes of Practice.

- Approved guidance (such as from the HSE or British Standards).

An example would be the standards required in respect of the use of mobile work equipment, such as a dumper truck. The **Provision and Use of Work Equipment Regulations 1998** have an Approved Code of Practice and guidance that details specific standards in respect of mobile work equipment.

Standards may be defined in legislation

Formulation and Prioritisation of Actions

When deciding on what action to take, you should always follow the hierarchy of controls:

- **Elimination** – can I remove the hazard altogether? If not, how can I control the risks so that harm is unlikely?

- **Substitute the hazard** – try a less risky option (e.g. switch to using a less hazardous substance).

- **Contain the risk** – prevent access to the hazard (e.g. by guarding).

- **Reduce exposure to the hazard** – reduce the number of persons exposed to the hazard and/or reduce the duration of exposure.

- **Personal protective equipment** – provide protection for each individual at risk.

- **Skill/supervision** – rely on the competence of the individual.

- **Welfare arrangements** – provide washing facilities to remove contamination and first-aid facilities.

Invariably, combinations of control are applied rather than relying on one alone. Just because a measure is near the bottom of the hierarchy does not mean it is not important (e.g. first aid) – it just means that an employer should not rely exclusively on it and must consider measures higher up the hierarchy.

If we have made an evaluation of the risk we should then be able to prioritise the necessary actions in terms of risks that need immediate attention, e.g. because of serious non-compliance, and those that may be dealt with in the short or even long term when resources become available. These actions may reflect going beyond the minimum legal standard and be best practice.

Recording Requirements

Clearly, it is good practice to record the details of risk assessment. If an employer has five or more employees, then it is a legal requirement to record the significant findings and any group of their employees identified by it as being especially at risk (**Regulation 3**, **Management of Health and Safety at Work Regulations 1999**).

The significant findings should include:

- A record of the preventive and protective measures in place to control the risks.

- What further action, if any, needs to be taken to reduce risk sufficiently.

- Proof that a suitable and sufficient assessment has been made.

In many cases, employers (and the self-employed) will also need to record sufficient detail of the assessment itself, so that they can demonstrate, say to an inspector, safety representatives or other employee representatives, that they have carried out a suitable and sufficient assessment. This record of the significant findings will also form a basis for a future revision of the assessment.

Use and Limitations of Generic, Specific and Dynamic Risk Assessments

- **Generic Risk Assessments**

  These risk assessments apply to commonly identified hazards and set out the associated control measures and precautions for that particular hazard. They give broad controls for general hazards but do not take into account the particular persons at risk or any special circumstances associated with the work activity. HSE guidance contains a wealth of information on hazards and controls required for a wide range of health and safety topics and can be used as the basis for generic risk assessments. In-house generic risk assessments can be used in workplaces where the particulars of the individuals at risk are not relevant and the activity is one that is standard and routine.

- **Specific Risk Assessments**

  These apply to a particular work activity and the persons associated with it. Specific activities, processes or substances used that could injure persons or harm their health are identified, along with exactly who might be harmed. Some workers have particular requirements such as new and young workers, migrant workers, new or expectant mothers, people with disabilities, temporary workers, contractors, homeworkers and lone workers. The risk assessment needs to be specifically tailored to the individuals at risk as well as the specific nature of the work task.

- **Dynamic Risk Assessments**

  Dynamic Risk Assessments (DRAs) are needed when work activities involve changing environments and individual workers need to make quick mental assessments to manage risks. Police, fire-fighters, teachers and lone workers, for example, often have to make swift risk judgments and identify controls, sometimes on their own and in high-pressure, potentially stressful, environments. To deal with these situations, DRAs are required. DRA is *"the continuous assessment of risk in the rapidly changing circumstances of an operational incident, in order to implement the control measures necessary to ensure an acceptable level of safety"*.

  Source: *Dynamic management of risk at operational incidents,* HMSO,1998

The Home Office's 1998 **Dynamic Risk Assessment Method** sets out five stages:

- **Evaluate the situation**: consider issues such as what operational intelligence is available, what tasks need to be carried out, what are the hazards, where are the risks, who is likely to be affected, what resources are available?

- **Select systems of work**: consider the possible systems of work and choose the most appropriate. The starting point must be procedures that have been agreed in pre-planning and training. Ensure that personnel are competent to carry out the tasks they have been allocated.

- **Assess the chosen systems of work**: are the risks proportional to the benefits? If yes, proceed with the tasks after ensuring that goals, both individual and team, are understood; responsibilities have been clearly allocated; and safety measures and procedures are understood. If no, continue as below.

– **Introduce additional controls**: reduce residual risks to an acceptable level, if possible by introducing additional control measures, such as specialist equipment or personal protective equipment.

– **Re-assess systems of work and additional control measures**: if risks remain, do the benefits from carrying out the task outweigh the costs if the risks are realised? If the benefits outweigh the risks, proceed with the task. If the risks outweigh the benefits, do not proceed with the task, but consider safe, viable alternatives.

'Suitable and Sufficient'

TOPIC FOCUS

Regulation 3, **Management of Health and Safety at Work Regulations 1999**, requires the risk assessment to be 'suitable and sufficient'. The term is not specifically defined in the Regulation but, to be effective, the risk assessment:

- Should identify the significant risks arising from, or in connection with, the work. The detail in an assessment should be proportionate to the risk so, if the hazards are simple, the assessment and record can be straightforward based on informed judgment and reference to relevant guidance.

- In many intermediate cases, the risk assessment will need to be more detailed and may need access to specialist guidance and the use of analytical techniques, e.g. a noise meter to measure noise levels.

- Will be the most sophisticated at the most hazardous site, especially where there are complex or unusual processes. For example, if a site stores bulk amounts of hazardous substances then the risk assessment may require the use of techniques such as quantified risk assessment (see later in the element).

- Must also consider all those who might be affected by the activities, whether they are workers or others, such as members of the public. For example, the assessment produced by a railway company will need to consider the hazards and risks which arise from the trains that not only affect employees but also contractors and the public.

- Should indicate the period of time for which it is likely to be valid. This will allow management to know when short-term control measures should be reviewed and modified.

Employers and the self-employed are required to take reasonable steps to help themselves identify risks by accessing appropriate legislation and guidance, manufacturers' instructions or seeking competent advice.

Limitations of Risk Assessment Processes

Risk assessment involves the evaluation of the likelihood of harm and its consequences for populations or individuals. Risk control requires the prioritisation of risks and the introduction of measures to prevent or reduce the harm from occurring. It is often assumed that an assessment of risks is scientific and objective whereas risk control is less straightforward because it combines the findings of risk assessment with other inputs, such as cost, risk perception, availability of technologies, etc. where there is more room for subjectivity. In practice, it is difficult to separate the two processes and assess risks without making assumptions. Consequently, risk assessment becomes a mixture of science and policy and a tool for extrapolating from statistical and scientific data to obtain a value which people will accept as an estimate of the risk attached to a particular activity or event.

There is also the public's attitude to acceptance of risk to consider. Nuclear power is regarded by some people to be too dangerous no matter how low the risks are. Other people question the premise on which risk assessment is based, which is that it is acceptable for certain persons to be exposed to particular risks so that others may benefit. In addition, there is some scepticism about the meaningfulness of low probability estimations for high-risk outcomes, with evidence quoted from major incidents such as Three Mile Island in the USA and Chernobyl in the Ukraine.

There are also issues regarding the general accuracy of risk estimations. One view is that risk assessment systematically overestimates risks by representing worst case scenarios and therefore causes unnecessary alarm and concern among the public. On the other hand, there is the belief that risk assessment may often underestimate the true magnitude of the problem, particularly in the case of health risks, by ignoring factors such as synergic exposures (the interaction of two agents producing a combined effect greater than the sum of their separate effects) or the variations in susceptibility among individuals.

It must therefore be accepted that assessing risks involves uncertainties, and that the science on which most risk assessment judgments are based is often inconclusive. Risk assessment relates to hypothetical rather than real persons and is inevitably based on value-laden assumptions. In view of these shortcomings, risk assessment continues to be a valuable tool for informing decisions but has serious limitations if used to blindly dictate them.

Temporary and Non-Routine Situations

The standard risk assessment aims to identify accurately all potential hazards, often by a walk around the workplace to assess activities, processes or substances used that could injure employees or harm their health. With familiarity, it is easy to overlook some hazards, so it is important to make sure that those that matter are identified. Manufacturers' instructions or data sheets for chemicals and equipment can be very helpful in explaining the hazards and putting them in their true perspective. Accident and ill-health records can help to identify the less obvious hazards, but it is also important to take account of **temporary** and **non-routine operations**. Maintenance may be routinely planned but it will be temporary so can be missed in a workplace walkabout. Emergency maintenance will be non-routine therefore needs to be assessed specifically. Cleaning operations are similar in that they may be routine but temporary or may result from emergency releases that are non-routine.

Consideration of Long-Term Hazards to Health

Another important issue easily missed by a standard approach to risk assessment is that of exposure to long-term hazards to health. Health hazards such as radiation, harmful substances and noise may not be readily observed in the workplace and need special consideration in the risk assessment process. Accident and ill-health records are unlikely to assist in identifying these types of hazard since the latency period before health effects are realised may be many years. Observation of the workplace is also ineffective since many health hazards are invisible and need specialist equipment to detect them.

Warning
Non ionising radiation

The starting point needs to be an accurate hazard profile of the workplace which will recognise and identify potential long-term hazards to health. Radiation, harmful substances and noise have their own specific risk assessment methodologies, detailed in the relevant legislation, which need to be followed in these circumstances.

Types of Risk Assessment

Before we continue, let's remind ourselves of the definitions of the following two terms:

- **Quantitative** – a measurement of magnitude is involved, e.g. there were four fatalities due to falls from height over a 12-month period at Business X; the airborne concentration of formaldehyde in a workplace was measured as 13ppm.

- **Qualitative** – no actual measurement is used. It involves describing the qualities, e.g. the airborne concentration was high or serious; the injury sustained was minor.

There are conceptually two basic categories of risk assessment: qualitative and quantitative. In practice, there is also a third category which uses numbers to indicate rank order, called semi-quantitative. Quantitative risk assessment uses more rigorous techniques in an attempt to quantify the magnitude of the risk. Even in the high-hazard industries (such as nuclear and chemical) most of the assessments are not quantitative. However, they are often used to satisfy a regulator that very unlikely events which, if they occurred, would have serious consequences not only to the organisation but also to the public (such as loss of containment of radioactive material in a nuclear facility) have been assessed. All risk assessments involve at least some element of subjectivity or judgment.

The definitions used below are based on those from HSG190 *Preparing safety reports*.

Qualitative Risk Assessments

Qualitative risk assessments are based entirely on judgment, opinion and experience, including approved guidance, rather than on measurements. They use technology-based criteria to establish if you have done enough to control risks, i.e. 'If I use this standard control measure I'm pretty sure the risk will be adequately controlled'. They allow you to easily prioritise risks for further action but while they enable risks to be ranked against other risks, they do not objectively estimate risks and so do not allow direct comparisons with external estimates.

A qualitative risk assessment is carried out by the risk assessor(s) making qualitative judgments with respect to the likelihood and consequence associated with a particular loss event. This judgment may be made through observation and discussion with employees as well as looking at other information, e.g. accident records. There are various ways in which likelihood and consequence could be categorised; the following is a simple example.

Example

Consider our torn carpet example again. There are a number of possible outcomes **should** someone trip on it; the severity categories might be:

- **Minor** – minor injury or illness with no significant lost time, such as a slight cut or bruise.
- **Lost time** – more serious injury causing short-term incapacity from work or illness causing short-term ill health, e.g. broken limb.
- **Major** – fatality or injury/illness causing long-term disability.

We use our experience to qualitatively judge the **most likely outcome**. We need to be sensible here, otherwise we will end up with the worst possible consequence always being death (or even multiple deaths) and the ability to prioritise remedial action is defeated as a result.

In this example, the likely outcome could be that someone would be badly bruised with no significant lost time and therefore 'Minor' would be chosen.

The likelihood of someone tripping over the carpet could be categorised by one of the following terms:

- Very likely.
- Likely.
- Unlikely.

In this example, we might judge that the likelihood of exposure to the hazard (coming into contact with the torn carpet) and subsequently tripping might be 'Likely'.

Think of our earlier equation for risk magnitude:

$$\text{Risk} = \text{Likelihood (or Frequency)} \times \text{Consequence (or Harm or Severity)}$$

It follows then, that although no numbers are being used, it is easy to see that the risk of someone injuring themselves on the torn carpet is moderate. For this reason, remedial action must be carried out to minimise the risk which in the short term might involve using carpet tape to join the two ends. Where there are a number of hazards that have been assessed in a similar way then it is possible to prioritise the remedial action so that the ones that pose the greatest risk are resolved first.

Clearly, each organisation would need to come up with their own categories which reflect the types of injury that may occur along with their likelihood frequency.

If the descriptors 'Minor', 'Lost time', 'Major' are replaced arbitrarily with '1', '2' and '3', respectively, this is still a qualitative risk assessment since there is still no quantitative basis for the choice – just a switch of numbers for words. So, don't think that the mere presence of numbers somehow converts it into a more thorough quantitative assessment.

Semi-Quantitative

In many risk assessments where the hazards are not few and simple, nor numerous and complex, it may be necessary to use some semi-quantitative assessments in addition to the simple qualitative assessments.

This may involve measuring the exposure of a worker to a hazardous substance or noise which can then be used to assess whether the risks to the workers are acceptable or not.

Semi-quantitative risk assessments may also use a simple matrix to combine estimates of likelihood and consequence in order to place risks in rank order as shown here in a simple 3 × 3 matrix.

Matrix ranking risk in terms of likelihood and consequence

Likelihood	**H = 3**	3	6	9
	M = 2	2	4	6
	L = 1	1	2	3
		Low = 1	**Medium = 2**	**High = 3**
			Consequence	

The likelihood and consequence are each characterised as low, medium or high and are assigned a number 1, 2 or 3, respectively. The risk is determined by calculating the product of the likelihood and the consequence, so risks range from 1 (low likelihood and low consequence) to 9 (high likelihood and high consequence).

The key point about such matrices is that they are used to rank risks, i.e. put them in order. They have no meaning in terms of their relative sizes so it cannot be assumed that a risk value of 9 is nine times the size of a risk rating of 1.

Quantitative Risk Assessment

Quantitative Risk Assessments (QRAs) attempt to calculate probabilities or frequencies of specific event scenarios. This is sometimes mandated by legislation, so that the results can be compared with criteria on what is considered an acceptable or a tolerable risk. They may use advanced simulation or modelling techniques to investigate possible accidents and will utilise plant component reliability data.

They are sometimes referred to as QRA or Probabilistic Risk Analysis (PRA).

DEFINITION

QUANTITATIVE RISK ASSESSMENT

"This is the application of methodology to produce a numerical representation of the frequency and extent of a specified level of exposure or harm, to specified people or the environment, from a specified activity. This will facilitate comparison of the results with specified criteria."

The degree of quantification used is variable. This type of risk assessment typically uses advanced tools such as Fault Tree Analysis (FTA) and Event Tree Analysis (ETA) (see later). It relies heavily on having suitable data to calculate the probability or frequency of a defined event.

QRAs are evidence-based (i.e. use 'hard' data) to be as objective as possible. It may not be possible to fully quantify risks – especially for infrequent events. Despite the name, QRAs invariably involve some subjectivity; this is because some broad assumptions may have to be made, e.g. in the application of human reliability assessment. This approach is used for safety cases to establish that the risks have been fully identified and to justify that enough has been done to reduce the risk to the lowest level reasonably practicable.

QRA is used in high-hazard chemical and nuclear installations and in the offshore oil industry for specific risk scenarios. They are included as part of their safety report requirements. Quantitative methods are also used in setting Workplace Exposure Limits (WELs) for airborne contaminants.

High-hazard installation

For major hazard sites, such as large chemical installations, numerical estimates of the probability or frequency of plant failure may be calculated. This is at its most valid when it involves the use of component reliability data, simply because the data is available (or can be measured) and is often based on a large sample size and so is statistically valid. The results of particular failure scenarios would then be considered, in terms of the different possible consequences, perhaps using fault trees and event trees (which may use component, structural, system and/or human reliability data). Consequence itself is not usually quantified as such. Rather, many failure scenarios (perhaps several hundred), all with different consequences are modelled and the probability or frequency of each scenario actually developing is calculated. An example would be the failure of a chlorine storage vessel in a particular way with dispersion modelling of several different release patterns for a toxic gas cloud. In such cases, the likelihood of harm resulting from all the different potential causes of failure has to be rolled up into a single estimate of the risk from that installation. We will consider some of these points in more detail when we look at failure tracing methods later in this element.

Organisational Arrangements for an Effective Risk Assessment Programme

As risk assessment is a fundamental component of a health and safety management system, it is important that the process of risk assessment is effectively managed. In an earlier element we looked at what constitutes an effective health and safety management system, including various models such as HSG65 *Managing for health and safety*. It is sensible to use such a model for managing risk assessment.

"*Plan*

* *... where you are now and where you need to be.*

* *... what you want to achieve, who will be responsible for what, how you will achieve your aims, and how you will measure your success. ... write down this policy and your plan to deliver it.*

- *Decide how you will measure performance... looking ... for leading ... and lagging indicators ... active and reactive ... .*

- *Remember to plan for changes and identify any specific legal requirements that apply ... .*

Do

- *Profiling your organisation's health and safety risks:*

  - *Assess the risks, identify what could cause harm in the workplace, who it could harm and how, and what you will do to manage the risk.*

  - *Decide what the priorities are and identify the biggest risks.*

- *Organising for health and safety.*

  In particular, aim to:

  - *Involve workers and communicate, so that everyone is clear on what is needed and can discuss issues – develop positive attitudes and behaviours.*

  - *Provide adequate resources, including competent advice where needed.*

- *Implementing your plan:*

  - *Decide on the preventive and protective measures needed and put them in place.*

  - *Provide the right tools and equipment to do the job and keep them maintained.*

  - *Train and instruct, to ensure everyone is competent to carry out their work.*

  - *Supervise to make sure that arrangements are followed.*

Check

- *Measuring performance:*

  - *Make sure that your plan has been implemented.*

  - *Assess how well the risks are being controlled and if you are achieving your aims. ...*

- *Investigate the causes of accidents, incidents or near misses.*

Act

- *Review your performance:*

  - *Learn from accidents and incidents, ill-health data, errors and relevant experience, including from other organisations.*

  - *Revisit plans, policy documents and risk assessments to see if they need updating.*

- *Take action on lessons learned:*

- *Include audit and inspection reports."*

> Source: HSG65 *Managing for health and safety,* HSE, 2013 (www.hse.gov.uk/pubns/priced/hsg65.pdf)

Acceptability/Tolerability of Risk

The criteria by which we, as a society, decide which risks we are prepared to expect workers and members of the public to live with, and those we are not, are set out in the HSE document *Reducing risks, protecting people* (R2P2).

Broadly speaking, we classify risks into three categories:

- **Acceptable** – no further action required. These risks would be considered by most to be insignificant or trivial and adequately controlled. They are of inherently low risk or can be readily controlled to a low level.

- **Unacceptable** – certain risks that cannot be justified (except in extraordinary circumstances) despite any benefits they might bring. Here we have to distinguish between those activities that we expect those at work to endure, and those we permit individuals to engage in through Tolerable their own free choice, e.g. certain dangerous sports/pastimes.

- **Tolerable** – risks that fall between the acceptable and unacceptable. Tolerability does not mean acceptable but means that we, as a society, are prepared to endure such risks because of the benefits they give and because further risk reduction is grossly out of proportion in terms of time, cost, etc. In other words, to make any significant risk reduction would require such great cost that it would be out of all proportion to the benefit achieved.

When we discuss **benefits** we mean:

- Employment.

- Lower costs of production.

- Convenience to the public.

- Maintenance of a social infrastructure, e.g. supply of food or production of electricity.

These three terms are best illustrated by the figure.

You can see that the risks that fall into the tolerable region are described as being 'As Low As is Reasonably Practicable', often referred to as 'ALARP'.

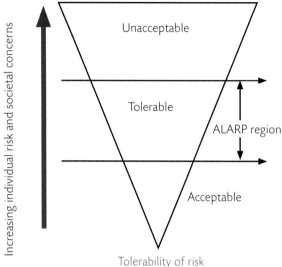

Tolerability of risk

These are risks that society is prepared to endure on the following assumptions:

- They are properly assessed to determine adequate control measures.

- The residual risk (after the implementation of control measures) is not unduly high.

- The risks are periodically reviewed to ensure they remain ALARP. How, then, are tolerability limits defined?

- **Boundary Between Acceptable and Tolerable**

  The HSE believes an individual risk of death of one in 1 million per year or less for both workers and members of the public is broadly acceptable. This risk is very low; indeed, using gas or electricity, or travelling by air poses a much greater risk.

- **Boundary Between Tolerable and Unacceptable**

  Here, there is a distinction between workers and the public:

  - For **workers**, the HSE proposes that an individual risk of death of one in 1,000 per year represents the dividing line between what

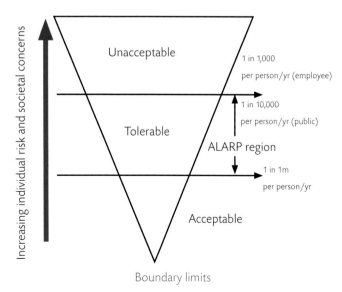

Boundary limits

is tolerable for an individual for any large part of their working life and what is unacceptable (apart from exceptional groups).

- For **members of the public** who have risks imposed on them who live, for example, next to a major accident hazard, the figure is an individual risk of death of one in 10,000 per year, i.e. 10 times less risk. This figure equates approximately to the individual risk of death per year as a result of a road traffic accident.

MORE...

You can access the HSE publication *Reducing risks, protecting people* (R2P2) from:

www.hse.gov.uk

You will also be able to find sample risk assessments at:

www.hse.gov.uk/risk/casestudies

STUDY QUESTIONS

5. What are the characteristics of a suitable and sufficient risk assessment?

6. Explain the following categories of risk:

   (a) Acceptable.

   (b) Unacceptable.

   (c) Tolerable.

(Suggested Answers are at the end.)

Systems Failures and System Reliability

IN THIS SECTION...

- Systems comprise a collection of interrelated processes that all need to be managed as a whole.

- Systems can have very complicated interactions between processes and the failure of the system (or potential failure) may need detailed investigation to discover the (potential) causes by adopting both holistic and reductionist approaches.

- The overall reliability of equipment depends on both the reliability of all components and the way in which they are arranged. As well as the individual reliability of each element, the system reliability will be affected by the way those elements are connected together. They may be connected:

  - In series, i.e. one after the other, so that the failure of any one piece means the failure of the system.

  - In parallel, i.e. side by side.

  - As a combination of both.

- Methods for improving system reliability include:

  - Use of reliable components.

  - Quality assurance.

  - Parallel redundancy.

  - Standby systems.

  - Minimising failures to danger.

  - Planned preventive maintenance.

  - Minimising human error.

Meaning of the Term 'System'

A system is:

- A group of interacting, interrelated, or interdependent elements forming a complex whole.

- A set or arrangement of things so related or connected as to form a unity or organic whole.

Systems comprise a collection of interrelated processes that all need to be managed as a whole. In health and safety, we are often interested in the organisation as a system so that, when things go wrong, we need to be able to analyse the system to find out which parts have failed.

The organisation is a system

Principles of System Failure Analysis

Systems can have very complicated interactions between processes and the failure of the system (or potential failure) may need detailed investigation to discover the (potential) causes. There are two basic approaches: holistic and reductionist.

Holistic Approach

This requires looking at the behaviour of the total system rather than the isolated workings of individual components, e.g. the workings of a car or the use of a telephone.

Holistic means trying to understand all the interactions between the separate components as they work together as a whole – everything affects everything else.

Reductionist Approach

This approach is when the system is divided into its components for individual analysis to identify system or subsystem failures, e.g. in a HAZOP or Failure Mode and Effects Analysis (FMEA) study (see below).

Analytical Considerations of Systems and Subsystems Failures

Given the need for a systems approach to risk management and that a thorough investigation of an accident, incident or disaster requires a detailed analysis of the underlying causes, we need to understand how complex systems such as organisations, process plant, items of equipment or human/machine interfaces can be broken down into sub-elements for more detailed investigation.

Failure tracing methods are a good example of how to treat the fault, failure or events systemically:

- A **HAZOP** is a powerful tool, developed primarily for use on chemical process plants but now with wide applicability. It employs a methodical approach using specialists guided by a formal system. The process critically examines sub-components of the process system (e.g. vessels, tanks, pipework) using guide words such as 'high, low, more, less' applied to key parameters such as pressure, temperature, flow, etc. The aim is to identify deviations from design intent that could have critical consequences and establish necessary safeguards at the design stage.

- **FMEA** is a simple but effective tool to improve reliability. The purpose of the analysis is to explore the effect of failures or malfunctions of individual components within a system. Consequently, the system needs to be broken down into sub-components which can then be analysed for failure. So, for each sub-component, we examine the possible failure modes, the effect of this failure, and the consequences of the failure in terms of severity and likelihood of detection, which can be allocated a Risk Priority Code (RPC). This analytical approach allows us to focus on the critical failure modes where we need to improve reliability.

ITEM	COMPONENT	FAILURE MODE	EFFECT	RPC	ACTION REQUIRED

Failure mode and effects analysis table

- **FTA** acknowledges the fact that most accidents are multi-causal, and employs analytical techniques to trace the events that could contribute. The fault tree is a logic diagram which traces all the branches of events that could contribute to an accident or failure. Consequently, we need to be able to identify the sub-elements that have a bearing on the final event; e.g. for an explosion, we need a flammable atmosphere, a source of ignition and enough oxygen. We then examine each of these sub-components to identify how they could arise. We can use quantified techniques, if necessary, to establish the critical events where reliability needs to be improved and introduce measures which will make the original accident or failure less likely.

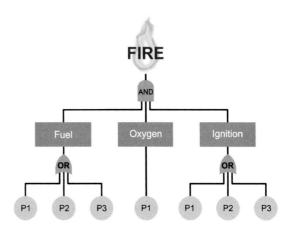

Simple fault tree for an explosion

- **ETA** starts with a primary event, then develops the resulting sequence of events that describe potential accidents, examining both the success and failure of safeguards as the accident sequence progresses. Event trees provide a methodical way of recording accident sequences and defining the relationships between initiating events and subsequent events within the system under study.

The selection of an appropriate tool for system and subsystem analysis will depend on the size and complexity of the system. It may be that a combination of tools is used, with HAZOPs or FMEA identifying critical failure modes, and ETA and/or FTA quantifying the consequences.

Using Calculations in the Assessment of Systems Reliability

Equipment Reliability

A question which is directly related to the maintenance and replacement of equipment concerns its reliability.

If equipment becomes unreliable and starts to break down, there will come a point when it is better to replace it than continue repairing and maintaining. It is, however, possible to increase the reliability of the process by having standby equipment in reserve, which is only used when there is a breakdown.

Some costs are increased (by having unused equipment), but this may be more than offset by the benefits from a more reliable process. This is the principle of the spare wheel in a car.

Thus, a reliability of 90% means there is a probability of 0.9 (out of 1) that the part will continue to operate normally for the period under investigation. To simplify things, we will phrase the discussion in terms of equipment made up of components, but this is not meant to imply any limit on applications.

The overall reliability of equipment depends on both the reliability of all components and the way in which they are arranged. If a single component has a reliability of R, putting two identical components in parallel will increase the overall reliability. The assumption is that the second component will only start to operate when the first one fails or is out of commission, and that the system can work adequately with only one of the components operating. Adding more components in parallel increases reliability, as the equipment will only fail when all components fail.

In many cases, a system consists of several individual elements or subsystems. Each element will have a reliability value of its own which contributes to the overall system reliability. As well as the individual reliability of each element, the system reliability will be affected by the way those elements are connected together. They may be connected:

- In series, i.e. one after the other, so that the failure of any one piece means the failure of the system.
- In parallel, i.e. side by side.
- As a combination of both, which is quite common.

By 'reliability' we mean the probability of functioning when required. So a system which has a reliability of 0.95 will operate 95 times out of 100, in the long term. If there were five individual elements connected in series, with each element having a reliability of 0.99, the overall system reliability would be $0.99^5 = 0.95099$ or 95.1%. If there were 10 such elements so connected, the overall reliability would be $0.99^{10} = 0.90438$ or 90.4%.

Consider a system of two identical components in parallel, with the reliability of each component R. The probability that a component continues normal operations is R, so the probability that it will stop operating during a specified period is $1 - R$. The probability that both components fail is $(1 - R)^2$. The reliability of the system is the probability that at least one of the components is operating, which is $1 - (1 - R)^2 - (1 - R)^2$. Similarly, the probability that n identical components in parallel will all fail is $(1 - R)^n$, and the reliability of the system is $1 - (1 - R)^n$. It follows that **any system of parallel components is more reliable than the individual components**.

If components are added in series, the reliability of the system is reduced. This is because a system with components in series only works if all separate components are working. Consider two components in series. If the reliability of each is R, the reliability of the two is the probability that both are working, which is R^2. If there are n components in series, their reliability is R^n. Thus, a system of components in series is less reliable than the individual components. These calculations are explained more fully in the following subsections.

Parallel Systems

In a parallel system, the failure of one component will not stop the system functioning.

The reliability of the system is described mathematically as:

$$R_{(S)} = 1 - [(1 - R_{(A)})(1 - R_{(B)})]$$

(You are not required to know how the mathematics for this works, merely to remember the equation.)

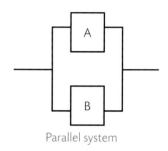

Parallel system

The number of terms will increase with an increasing number of components in parallel.

This applies no matter how many components are in the system. For example, consider the next figure:

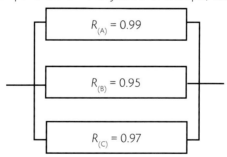

$R_{(A)} = 0.99$

$R_{(B)} = 0.95$

$R_{(C)} = 0.97$

Three components in parallel

The formula for this would be:

$$R_{(S)} = 1 - [(1 - R_{(A)})(1 - R_{(B)})(1 - R_{(C)})]$$

$$R_{(S)} = 1 - [(1 - 0.99)(1 - 0.95)(1 - 0.97)]$$

$$R_{(S)} = 1 - [(0.01)(0.05)(0.03)]$$

$$R_{(S)} = 1 - [0.000015]$$

$$R_{(S)} = \textbf{0.999985 or 99.9985\%}$$

Notice how much change has been introduced to the system. In series, the reliability is reduced to less than any of the individual components; in parallel, it is increased.

In an attempt to improve the reliability, the possibility of having components in parallel throughout the system may be considered. Unfortunately, there would be a financial cost to this. All the additional components that would have to be included would add to the cost of the finished product, with the result that it would be uneconomic to produce. There would also be a subsequent increase in size to accommodate the extra components.

Series Systems

In series, components are joined to each other such that all must function for the system to operate. The following figure shows two components in series:

To calculate the reliability of the series system, the reliabilities are multiplied together:

$$R_{(S)} = R_{(A)} \times R_{(B)}$$

This applies no matter how many components are in the system. For example, consider the next figure:

$$\boxed{R_{(A)} = 0.99} \quad \boxed{R_{(B)} = 0.95} \quad \boxed{R_{(C)} = 0.97}$$

Three components in series

The reliability of the system $R_{(S)}$ is described mathematically as:

$$R_{(S)} = R_{(A)} \times R_{(B)} \times R_{(C)}$$

$$R_{(S)} = 0.99 \times 0.95 \times 0.97$$

$$R_{(S)} = \textbf{0.91 or 91\%}$$

Notice that the reliability figures are presented in the calculation as a figure, not a percentage. Also note how the individual effects quickly combine to reduce the reliability of a series system.

Mixed Systems

Unfortunately, systems are not composed solely of series systems or parallel ones but are generally mixed. To calculate the efficiency of a system, consider the example given in the following figure:

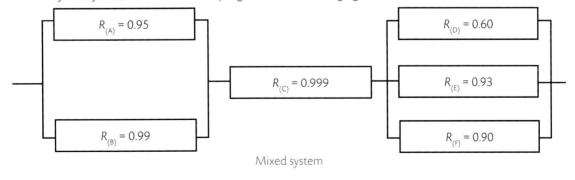

Mixed system

The basic principle is to break down the overall system into component series and parallel systems and treat each separately.

Looking at the parallel system involving $R_{(A)}$ and $R_{(B)}$:

$R_{(1)} = 1 - [(1 - R_{(A)})(1 - R_{(B)})]$

$R_{(1)} = 1 - [(1 - 0.95)(1 - 0.99)]$

$R_{(1)} = 1 - [(0.05)(0.01)]$

$R_{(1)} = 1 - (0.0005)$

$R_{(1)} = \textbf{0.9995}$

Looking at $R_{(D)}, R_{(E)}$ and $R_{(F)}$:

$R_{(2)} = 1 - [(1 - R_{(D)})(1 - R_{(E)})(1 - R_{(F)})]$

$R_{(2)} = 1 - [(1 - 0.60)(1 - 0.93)(1 - 0.90)]$

$R_{(2)} = 1 - [(0.40)(0.07)(0.10)]$

$R_{(2)} = 1 - (0.0028)$

$R_{(2)} = \textbf{0.9972}$

Thus the system can now be reproduced as a series system, shown in the following figure using the figures obtained for $R_{(1)}$ and $R_{(2)}$:

$$\boxed{R_{(1)} = 0.9995} - \boxed{R_{(C)} = 0.999} - \boxed{R_{(2)} = 0.9972}$$

Interim stage

$R_{(S)} = R_{(1)} \times R_{(C)} \times R_{(2)}$

$R_{(S)} = 0.9995 \times 0.999 \times 0.9972$

$R_{(S)} = \textbf{0.9957}$

This example shows what is done in industry. The components that are less reliable are put in a parallel system to increase their reliability, while those with good reliability are left in series.

Unfortunately, not all systems are designed to break themselves down into neat parallel or series packages. Sometimes the system will have cross connections.

To resolve the reliability of these systems, there are mathematical models that need to be used to analyse the system in each of the various operational modes and combine the findings. While these are important tools in the study of reliability, the mathematics and methods are beyond the scope of this course.

Common Mode Failures

Failure is defined as 'the termination of the ability of an item to perform a required function'. Common mode failure is where two or more components fail in the same way or mode due to a single event or cause, e.g. two or more pairs of braces attached to the **same** buttons will fail in the same way if the buttons fail. This will not happen if we have one pair of braces and a belt instead. Another example might be a machine where all the components are badly serviced by the same worker with poorly calibrated equipment. Yet another example is where several components are all connected to one other component – if that fails they all fail in the same way due to that cause.

Principles of Human Reliability Analysis

Think about a person who is driving an unfamiliar car. The driver knows how to drive (a licence confirms that, i.e. training and certification). The

HRA is a structured way of estimating the probability of human errors

driver will identify where all the controls are and what they do (familiarisation) and set off. A problem may occur if the windscreen wiper and indicator stalks are switched so that they are the opposite of what the driver is used to. Initially, indicating during driving will be correct, but at some point the windscreen wipers will operate instead of the indicators. A human error will have occurred in a situation where the driver has already demonstrated an understanding of the working of the indicator controls.

Hardware design can only go so far for improved reliability; there still exists the human input into the operation. We have only mentioned human reliability as basically 'unreliable'. The question we have to ask of this human input is: 'How unreliable is it?'. We then have to establish the answer to this question.

Humans do not work in the same way as machines. They are not good at carrying out repetitive tasks to a consistent standard and no two humans perform in the same way. The reliability of a human being cannot, therefore, be determined to the same accuracy as a machine, but action can be taken to make reasonable assessments of the type and frequency of error so that positive action can be taken to minimise the effects.

Human Reliability Analysis/Assessment (HRA) is a structured way of estimating the probability of human errors in specific tasks. It is used as part of certain risk assessment processes (e.g. QRA in the nuclear, offshore and chemical industries).

The methodology for HRA is similar in principle to task analysis:

- Determine scope of assessment (aim, tasks, etc.).

- Gather information (observation, etc.).

- Describe the tasks (goals, steps, interactions between person and system).

- Identify any potential human errors.

- Estimate overall Human Error Probabilities (HEPs) for the task (if needed): measure, calculate, use of experts, use of some formal methods (e.g. THERP, SLIM, HEART*). This area is based on some judgment – it is not precise and involves estimates.

- Give result to system analyst to incorporate into the overall risk assessment of the system and consider if human error has a significant impact on the system.

- Develop control measures (if significant risk).

*THERP is 'Technique for Human Error-Rate Prediction', SLIM is 'Success Likelihood Index Method', HEART is 'Human Error Assessment Reduction Technique'. HEART is a technique to arrive at HEPs by matching the task being assessed to one of nine generic task descriptions from a given database and then to modify the HEPs according to the presence and strength of the identified Error-Producing Conditions (EPCs).

Methods for Improving System Reliability

For any organisation, it is extremely important to have reliable systems in place to ensure that orders can be produced on time, downtime is kept to a minimum and, where reliability affects safety, to protect individuals. It is vital that reliability is designed in at every stage of the process.

Use of Reliable Components

A system is only as reliable as the components that make it up. For this reason, it is vital that suitable, good quality, well-proven components from reputable suppliers are used. It is important that quality checks are carried out on the parts to ensure that they meet legal specifications as well as any additional specified ones. Suppliers can be asked to provide details of their quality assurance procedures and testing regimes.

Quality Assurance

Materials will be delivered to the factory for processing into the finished product. As there are a number of opportunities for the product to fail to meet the required standard during the manufacturing process, there is a need to check at each stage. These checks should be recorded and a management process introduced that does that. This is **quality control**. The system will probably be one based on the BS ISO 9000 series of documents detailing **quality assurance**. These records are important in the event of a failure.

Parallel Redundancy

Additional components can be added in parallel series so that if one component fails, the other one will keep the system going. While this can be costly if components are expensive, it does mean that the system is less likely to fail as often and unplanned downtime is kept to a minimum.

Standby Systems

In order to prevent a system failure, a standby system can be installed so that, should part of the system or a component stop working, then an alternative system automatically steps in to continue operation. This type of system is invaluable where failure of the system could affect safety, e.g. lighting in an operating theatre.

Minimising Failures to Danger

When a system does fail, it is important that the failure does not end with the production of a hazardous situation. For this reason, it is vital that systems fail to safety. There are a number of ways of achieving this. One of the most important ways is through good design, e.g. ensuring that dangerous machinery has an automatic power cut-out as soon as a hazardous component fails.

Planned Preventive Maintenance

Planned preventive maintenance will improve safety and plant integrity as well as reliability. It is a means of detecting and dealing with problems before a breakdown occurs. For example, car manufacturers recommend that the oil is changed at specified intervals to prevent failure of the system and increase reliability.

Minimising Human Error

Human error does occur but can be minimised by ensuring that the:

- 'Right' person is doing the 'right' job.

- Individual has adequate training and instruction.

- Individual receives appropriate rest breaks.

- Worker-machine interface is ergonomically suitable.

- Working environment is comfortable, e.g. noise, lighting, heating, etc.

> STUDY QUESTIONS
>
> 7. (a) Define the term 'failure'.
>   (b) Explain what is meant by 'common mode failure'.
> 8. Outline the methods available for improving human reliability.
> 9. (a) Draw a diagram of a system containing two components connected in series.
>   (b) If the reliability of each component is 0.9, calculate the reliability of the complete system.
>
> (Suggested Answers are at the end.)

Failure Tracing Methodologies

IN THIS SECTION...

Failure tracing methods are structured techniques to assist in hazard identification and risk assessment. They include:

- Hazard and Operability Studies (HAZOPs) – identify hazards in a system and their effect on the system.

- Fault Tree Analysis (FTA) – identifies the necessary events and how they combine to lead to a loss event called the Top Event.

- Event Tree Analysis (ETA) – used to identify the possible consequences from an event and the influence of controls.

A Guide to Basic Probability

To understand the advanced risk assessment techniques that involve quantified risk assessment, you need to understand the basic principles of probability.

Probability of a Single Event Occurring

Probability relates to the chance of an event occurring and in numerical terms can only have a value between 0 and 1:

- 0 means there is no chance of it happening, i.e. it is impossible.

- 1 means it is certain to happen.

Suppose we toss a coin.

There are only two possible outcomes, namely heads or tails:

Probability is the chance of an event occurring

- So, the probability of getting heads, i.e. ½ or 0.5.

- The probability of getting tails is also ½ or 0.5.

You will notice that given there are only two possible outcomes, the sum of the two probabilities equals one, i.e. Probability of Heads + Probability of Tails = 0.5 + 0.5 = 1.

Similarly, if we threw a six-sided dice, the probability it would land showing a six would be ⅙ = 0.167.

Probability of Multiple Events Occurring

Suppose we now toss a coin twice. What is the probability that on both occasions it will show heads?

Here are all the possible outcomes:

- Heads and then heads.

- Heads and then tails.

- Tails and then heads.

- Tails and then tails.

There are four outcomes but only one matches 'Heads and then heads'; the probability is therefore ¼ = 0.25.

The probability is simply the product of each of the two events, i.e. ½ × ½ = ¼ = 0.25.

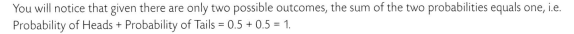

So if we wanted to know what the probability is of obtaining three sixes if we threw three separate dice, it would be: $\frac{1}{6} \times \frac{1}{6} \times \frac{1}{6} = \frac{1}{216} = 0.0047$, which is not very likely!

So to calculate the probability of two or more independent events occurring we **multiply** the probabilities.

Incidentally, the probability of winning the Lotto, i.e. selecting all six winning numbers, is 1 in 44,000,000, which is an extremely remote event! Yet, of course, as millions of people play every week, there is often a winner.

Suppose we now wanted to know what the probability is of getting a 1 OR a 6 if we threw a dice.

The possible outcomes are, of course, 1, 2, 3, 4, 5 and 6. Of these six possibilities, two meet our requirement of 1 or 6, i.e. 2 out of 6 = $\frac{2}{6}$ (or 0.33).

- The probability of getting a 1 is $\frac{1}{6}$.

- The probability of getting a 6 is also $\frac{1}{6}$.

So the probability of getting a 1 OR a 6 is $\frac{1}{6} + \frac{1}{6} = \frac{1}{3}$.

In other words, when we want to know the probability of one event or another, we ADD the probabilities of the two separate events.

Understanding these principles will be very useful to you when we look at FTA.

The probability of rolling a six is $\frac{1}{6}$

Probability and Frequency

In risk assessment, we sometimes refer to the probability of an event and sometimes to the frequency of an event. What is the difference?

- **Probability** is the chance something will happen. So, if the probability of tripping over a torn carpet happened to be 1 in a 1,000, then this means that if 1,000 people walked over the tear then, on average, one would trip. How long this would take would depend how long it takes 1,000 people to encounter the tear. If the tear was in a busy walkway then it might be only a matter of a few hours or less; if it was in a room that was hardly used then it might take many years. So exposure to hazard is very important.

- **Frequency** takes account of the exposure, so if we say the event will happen on average once every 10 years we say its frequency is $1/10$ or $0.1y^{-1}$.

   We cannot combine frequencies, i.e. add or multiply them in the same way that we can probabilities, but we can multiply a frequency with a probability.

   Consider an event which has a probability of harm of 1 in 100, i.e. 0.01, and we know the event occurs 300 times a year; the frequency of the harm will be:

   0.01 (probability of harm) × 300 (frequency of occurrence per year) = 3 harmful events a year.

Principles and Techniques of Failure Tracing Methods in the Assessment of Risk

Failure tracing methods may be used in more detailed risk assessments. They are unnecessary in most cases but provide a systematic methodology for identification of hazards and, in some cases, calculation of failure probabilities, for more complex cases; they are used extensively in, for example, quantified risk assessment. They can be used qualitatively and some quantitatively as well. Some can be used to model incidents and so can be used in accident investigation.

Hazard and Operability Studies

The HAZOP method is designed for dealing with relatively complex systems, such as large chemical plants or a nuclear power station, where a deviation from what is expected in one component of the system may have serious consequences for other parts of the system. The principles can be used in simpler situations, but a full HAZOP will not usually be cost-effective except in a high-risk situation. We will cover the principles and outline the technique.

HAZOPs are a form of structured 'brainstorming'.

The studies are carried out by a multi-disciplinary team, usually made up of four to seven people. Members typically include:

- Study leader – should not be closely involved with the project but should have good experience of HAZOPs and can keep the team focused.

- Recorder – documents the proceedings.

- Designer – explains how the system should work.

- A user of the system, such as a production manager.

Others who may contribute include:

- Maintenance engineer.

- Software specialist.

- Safety expert.

- Instrument engineer.

HAZOPs are carried out by multi-disciplinary teams

The team has a diagram that represents the system, showing each of the components of the system and how they are related. In the case of a chemical process plant, this would be a pipeline and instrumentation diagram.

The steps involved in a HAZOP are:

- Divide the system into parts (sometimes called 'nodes'). In the case of a chemical process plant, this might be a pipeline between a storage vessel and a reactor.

- For each part, define the design intention, i.e. what is meant to occur when it is operating normally.

- Apply a number of 'guide words' to the statement of intention, so that every possible deviation from the required intention is considered. The main guide words are:

NO or NOT	Negation of intention, e.g. no flow.
MORE	Quantitative increase, e.g. high pressure.
LESS	Quantitative decrease, e.g. low temperature.
AS WELL AS	Qualitative increase, e.g. impurity present.
PART OF	Qualitative decrease, e.g. only one of two components present.
REVERSE	Logical opposite of intention, e.g. backflow.
OTHER THAN	Complete substitution, e.g. flow of wrong material.

Let us consider a simple example.

In a batch process, two substances, A and B, are pumped from their respective storage vessels into a reaction vessel:

Parameter	Guide Word	Deviation	Cause	Consequence	Actions
Flow of 'A'	NO	No flow	Pump failure Pump off Line blockage Tank empty Reactor full	Incorrect product/ reaction doesn't occur	Indicate pump working at control panel Maintain lines Level control on tank, etc.
	MORE	More flow	Pump at wrong rate	Incorrect product/ reaction doesn't occur	Automatic control of pump rate

The design intention is for equal amounts of A and B to be pumped into the reactor vessel. We can identify two parts to this system:

- Storage vessel A and the pipeline and pump to the reactor vessel.
- Storage vessel B and the pipeline and pump to the reactor vessel.

Let us apply the first guide word 'NO' to the first part. In this case, it would mean 'no flow'. Think of some reasons why there might be no flow in the pipeline.

Here are some suggestions:

- Pump A has failed.
- Pump A is not switched on.
- Storage vessel A is empty.
- Reactor vessel is full.
- Pipeline is blocked.

Having established the possible causes of this deviation we then need to make an estimate of the risk for each cause. In other words:

- How likely is this deviation?
- How soon would we know that the deviation had occurred?
- What are the consequences?
- How serious are the consequences?

Having assessed the risk, we need to determine whether the existing control measures are adequate or whether we need to introduce additional ones. In our example, we might introduce a level gauge in the storage tank and reactor vessel, each linked to an alarm so that the operator of the plant would know when the levels were too high or too low. We might introduce a more reliable pump or even a second (redundant) pump that could take over should the first one fail.

Having considered this deviation, we would then move on to the next guide word, which in this case would mean 'more flow'. Having examined flow, we would move on to other parameters, such as temperature and pressure, and apply the guide words again.

Clearly, even a relatively simple system can result in a significant and lengthy analysis. For a major plant, it can take considerable time and involve significant expenditure.

Fault Tree Analysis and Event Tree Analysis

Be careful not to get confused between these two techniques; they are, in fact, complementary (and are often used together) but focus on opposite sides of an undesired event. The following figure shows how they fit together:

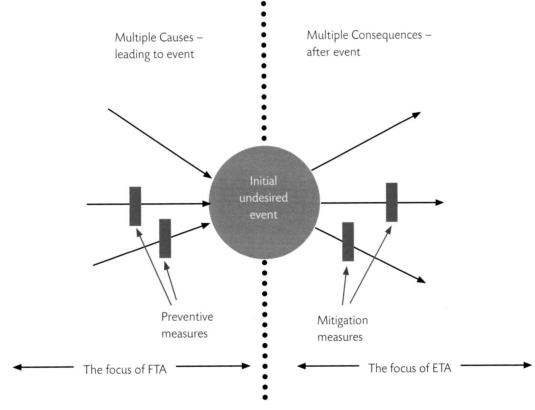

Relationship between FTA and ETA ('Bowtie' model)

The figure – known as the 'Bowtie' model, or 'Bowtie' analysis – only shows a single 'undesired event'; in reality, multiple causes can lead to many different events initially, each then escalating with multiple consequences. You can analyse each event with FTA and ETA. In summary, FTA is concerned with analysing faults which might lead to an event, whereas ETA considers the possible consequences once an undesired event has taken place. Both can be applied qualitatively or, if you have the data, quantitatively.

Fault Tree Analysis

In many cases, there are multiple causes for an accident or other loss-making event. FTA is an analytical technique used to trace the events which could contribute to the incident. It can be used in accident investigation and in a detailed risk assessment.

The fault tree is a logic diagram based on the principle of multi-causality, which traces all branches of events which could contribute to an accident or failure. It uses sets of symbols, labels and identifiers. For our purposes, you will only need a handful of these:

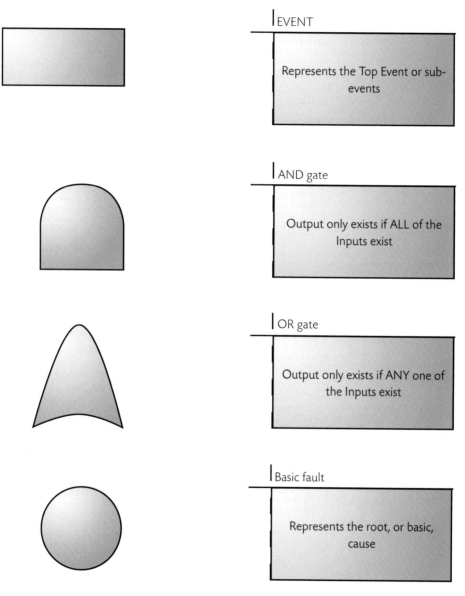

| EVENT
Represents the Top Event or sub-events

| AND gate
Output only exists if ALL of the Inputs exist

| OR gate
Output only exists if ANY one of the Inputs exist

| Basic fault
Represents the root, or basic, cause

A fault tree diagram is drawn from the top down (like an upside-down tree). The starting point is the undesired event of interest (called the Top Event because it gets placed at the top of the diagram). You then have to logically work out (and draw) the immediate and necessary contributory fault conditions leading to that event. These may each in turn be caused by other faults and so on. Each branch of the tree is further developed until a primary failure (such as a root cause) is identified. It could be endless (though, in fact, you will naturally have to stop when you get as far as primary failures). The most difficult part is actually getting the sequence of failure dependencies worked out in the first place. Let's look at a simple example of a fire to illustrate the point.

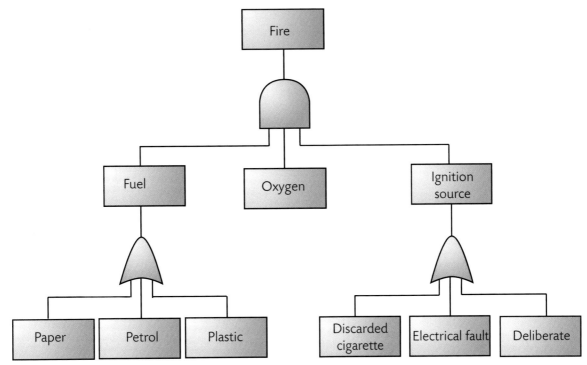

FTA of a fire

The above figure shows a simple fault tree for a fire. For the fire to occur, there needs to be:

- Fuel.
- Oxygen.
- An ignition source.

Notice we use an AND gate to connect them here because all three need to be present **at the same time** to allow the Top Event. The example shows that, in this scenario, there happens to be three possible sources of fuel and three possible sources of ignition. An OR situation applies in each case, because it would only need one of these to be present. The example also shows a single source of oxygen (e.g. the atmosphere).

HINTS AND TIPS

Don't worry about getting the symbols precisely right when you draw fault trees by hand; you can make your intentions quite clear by writing 'AND' or 'OR' in the appropriate logic gate as well. Also, as long as you describe the fault/failure in a box, don't worry too much about the (sometimes subtle) distinction between what should go in rectangles and circles.

To prevent the loss taking place, we would first examine the diagram for AND gates, because the loss can be prevented if just **one** of the conditions is prevented.

Fault trees can also be quantified but need relevant data on the respective probabilities of each of the sub-events. Let's try this on the same example.

From previous experience, or as an estimation, a probability for each of the primary failures being present or occurring can be established, shown in the following figure (these are purely illustrative):

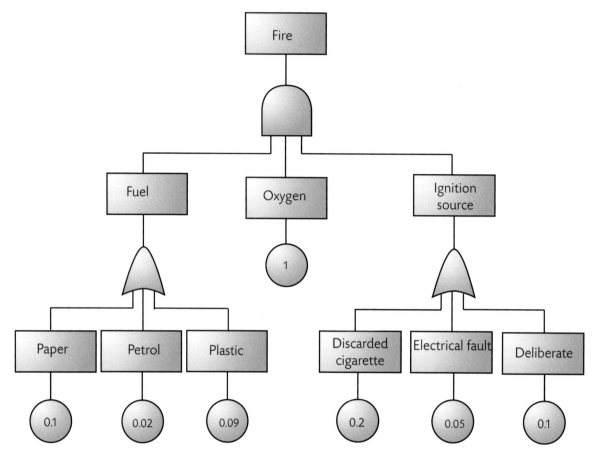

(We've assigned a probability of 1 for oxygen being present, as it is always in the surrounding atmosphere.)

We can then use two well-established rules of combination of these probabilities and progress up the diagram to get at the probability of the Top Event (fire) occurring. Essentially, we:

- **Add** the probabilities that sit below an OR gate (this isn't strictly correct, but is a 'rare event' approximation).

- **Multiply** the probabilities that sit below an AND gate.

So, in this example, combining probabilities upwards to the next level gives:

- Probability of FUEL being present = 0.1 + 0.02 + 0.09 = 0.21

- Probability of OXYGEN being present = 1

- Probability of IGNITION being present = 0.2 + 0.05 + 0.1 = 0.35

Updating the figure:

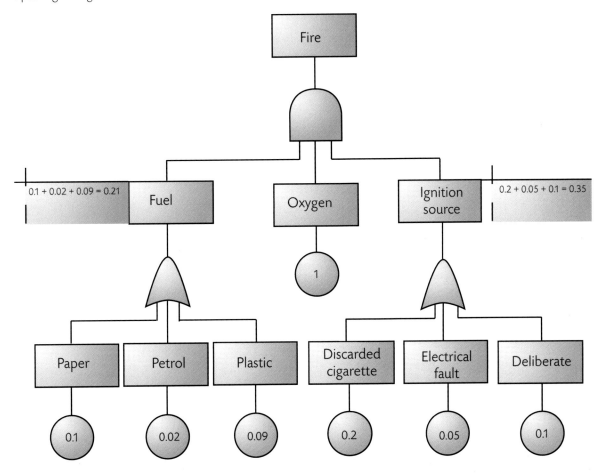

Moving up again, we can now calculate the probability of the Top Event. These faults are below an AND gate, so we multiply the probabilities, giving 0.21 × 1 × 0.35 = 0.0735. The fully quantified fault tree then looks like this:

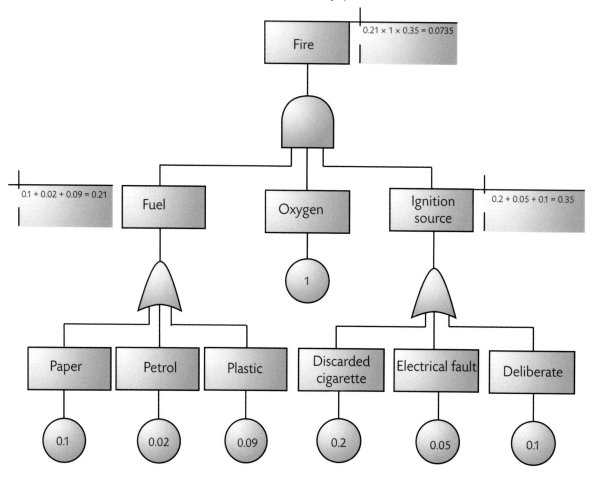

So the probability of the Top Event is 0.0735.

You are probably wondering what this number means.

Well, if the probability was 0.1, this would mean there was a 1 in 10 chance of it occurring (i.e. 1/0.1 = 10).

If it was, say, 0.25 then this represents 1/0.25 = 4, i.e. a 1 in 4 chance, so 0.0735 means 1/0.0735 = 13.6, i.e. nearly a 1 in 14 chance of the fire occurring.

HINTS AND TIPS

To get maximum marks in the exam, make sure that you show all your workings out when quantifying a fault tree.

Event Tree Analysis

Unlike identifying the root causes of an event under consideration, ETA is concerned with identifying and evaluating the consequences **following** the event.

In FTA, the main event is called the Top Event, whereas in ETA it is called the Initiating Event.

Event trees are used to investigate the consequences of loss-making events in order to find ways of mitigating, rather than preventing, losses.

The stages involved in carrying out an ETA are:

- Identify the Initiating Event of concern.

- Identify the controls that are assigned to deal with the Initiating Event, such as automatic safety systems, and other factors that may influence the outcome, such as wind direction or presence of an ignition source that would be important if there was an escape of a large amount of liquefied petroleum gas.

- Construct the event tree beginning with the Initiating Event and proceeding through the presence of conditions that may exacerbate or mitigate the outcome.

- Establish the resulting loss event sequences.

- Identify the critical failures that need to be dealt with.

- Quantify the tree if data is available to identify the likelihood or frequency of each possible outcome.

There are a number of ways to construct an event tree. They typically use binary logic gates, i.e. a gate that has only two options, such as success/failure, yes/no, on/off. They tend to start on the left with the Initiating Event and progress to the right, branching progressively. Each branching point is called a node. Simple event trees tend to be presented at a system level, glossing over the details.

Let us illustrate the process with a simple example where the Initiating Event is a fire:

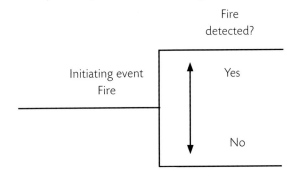

The ETA diagram shows the Initiating Event on the left, leading to a mitigation measure: 'Is the fire detected?' The answer to this question is a simple YES or NO, so the tree now branches to represent whether the answer to the question is YES or NO. Detection of the fire is, of course, the first step in minimising the consequence of the fire so now we need to consider those other factors that are necessary that will either minimise the outcome or make the situation worse.

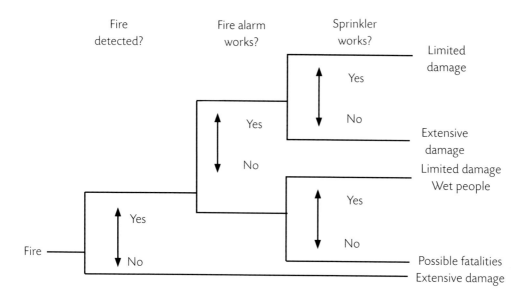

In this example, apart from detecting the fire, we need to have an effective alarm system to alert persons in the building and an effective sprinkler system. The detection system switches on the alarm and then the sprinkler. Each factor is considered as to whether it occurs or does not occur, each leading to a further branch in the tree.

You will also note that in this example, should the fire **not** be detected, the alarm will not sound and the sprinkler will not operate, so if the fire is not detected there will be the worst outcome.

At the end of the branches on the right of the tree, you can see the different outcomes identified depending on the success or otherwise of the intervening factors.

To **quantify an event tree** we need to know the probability for each of the outcomes that follow from the Initiating Event.

The probability that the:

- fire is detected;

- alarm works; and

- sprinkler works.

In binary logic, an event either happens or does not. Let us suppose that the probability the fire is detected is 0.95; this means 95 out of every 100 times a fire takes place it will be detected. It follows that the probability it is **not** detected is 0.05, i.e. 5 out of every 100 times.

So the:

Probability of success + Probability of failure = 0.95 + 0.05 = 1

In all binary events, the sum of the two probabilities must always equal 1.

Here is the tree with the probabilities for the three events after the Initiating Event.

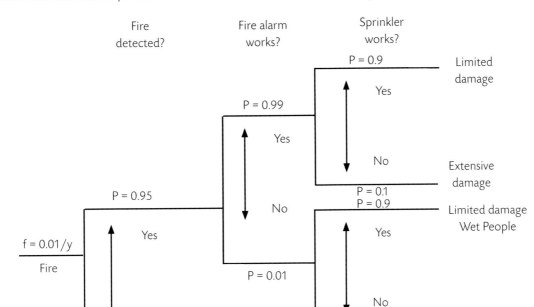

As we noted above, the sum of the probabilities for each event is always 1.

You will also notice that we have included the frequency of the Initiating Event, i.e. the fire, which is 0.01/y which means a fire will occur on average once every 1/0.01 years, i.e. once every 100 years.

We can now calculate the frequency of each of the possible sequences identified in the tree by multiplying the probabilities for each sequence and then multiplying this by the frequency of the Initiating Event.

So, for the sequence leading to 'Limited damage' we have:

Frequency of fire (0.01/y) × Fire detected (0.95) × Fire alarm works (0.99) × Sprinkler works (0.9) = 0.0085/y. This means this outcome will occur (on average) once every 1/0.0085 years = once every 118 years.

In contrast, the sequence in which the fire is **not** detected will occur with a frequency of 0.01/y × 0.05 = 0.0005/y which is once every 1/0.0005 years = 2,000 years.

Here is the tree with the expected frequencies for each of the outcomes:

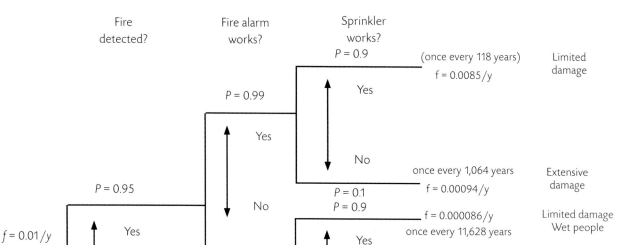

What can we conclude from this?

Well, the most frequent outcome, should a fire break out, is that all the controls will work and damage will be limited (frequency = 0.0085/y). However, the second most frequent outcome is the fire not being detected and this leads to extensive damage and possible fatalities (frequency = 0.0005/y).

What would we recommend?

One solution would be to increase the reliability of the fire detection system. However, this is a crucial link in the sequence, since if it fails then the reliability of the alarm and sprinkler become irrelevant. A much better improvement would be to include a second detector, independent of the first. This would mean that BOTH detectors would have to fail, which is a much less likely event.

The following is a similar example. The figure shows a quantified event tree for the action following a fire on a conveyor system. Here the fire detector, i.e. the heat sensor, opens the valve leading to operation of the water sprays. As in the previous example, should the sensor fail, the success of the valve or water spray is not relevant to the outcome; but here, should the valve fail, the success of the water spray becomes irrelevant.

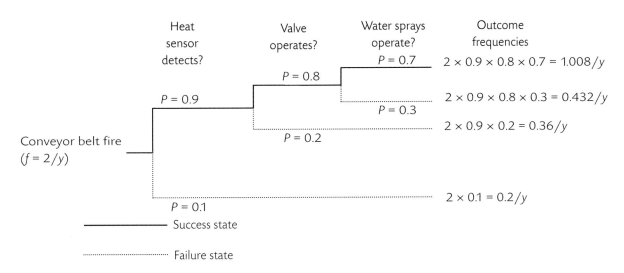

The only outcome resulting in control of the event is where the sensor, valve and water spray operate (the example is a little contrived but serves to demonstrate the principles).

Notice how the frequencies of the outcomes are calculated. Notice also that the sum of all the outcome frequencies adds up to 2 in this case, i.e. the frequency of the Initiating Event (the conveyor belt fire).

The event tree could be used to check that there were adequate fire detection, warning and extinguishing systems.

STUDY QUESTIONS

10. The frequency of pipework failure in a large liquefied petroleum gas storage facility is estimated at once every 100 years (f = 0.01/y). Immediate ignition of the released gas (probability, P = 0.05) will result in a jet flame. Otherwise, prevailing winds will normally carry any vapour cloud off site across open countryside where it will disperse safely. However, under certain conditions (P = 0.1), the cloud may drift to a nearby industrial estate where ignition (P = 0.5) will cause a vapour cloud explosion or flash fire.

11. Using the data provided, construct an event tree to calculate the expected frequency of fire or explosion due to pipework failure BOTH on site AND on the industrial estate.

12. Outline the basic principles of a HAZOP.

13. Briefly explain the difference between a 'fault tree' and an 'event tree'.

(Suggested Answers are at the end.)

Summary

Sources of Information Used in Identifying Hazards and Assessing Risk

We considered commonly used accident and disease ratios – incidence rate, frequency rate, severity rate and prevalence rate.

Information can be sourced both externally and internally:

- **External sources** include: the HSE, industry sources, trade unions, insurance companies, the ILO, the WHO, professional bodies and the EU.

- **Internal sources** include: accident reports, absence records and maintenance records.

Hazard Identification Techniques

Various techniques can be used to detect hazards, including: task analysis, checklists, observations and incident reports.

A **checklist**, which lists the key issues to be monitored, is developed to ensure a consistent and comprehensive approach is used to check that all the safety elements will be covered during an inspection.

The analyst should make an **observation** of the work being done, including work undertaken by groups of operators.

Assessment and Evaluation of Risk

We looked at the key steps that characterise all risk assessments:

- Hazard identification.

- Identify who is at risk.

- Estimate, evaluate risk and identify precautions.

- Record significant findings and implement.

- Review the assessment.

Assessments should be **suitable and sufficient**. This means that:

- They should identify the significant hazards.

- The detail should be proportionate to the risk with the most hazardous operations requiring the most sophisticated assessments.

- In some cases, more analytical techniques may be necessary (e.g. high noise levels).

- The period for which it is likely to remain valid should be identified.

Types of risk assessment include:

- Generic – apply to commonly identified hazards and set out the associated control measures and precautions.

- Specific – apply to a particular work activity and the persons associated with it.

- Dynamic – apply to work activities that involve changing environments and require quick mental assessments to manage risks.

- Qualitative – risks are represented by simple word descriptors.

  - **Risk** = probability (or frequency) × consequence (or harm or severity).

  - **Probability** is the chance that a given event will take place.

  - **Severity** of risk is the outcome.

- Semi-quantitative – results represented by qualitative and quantitative descriptions. In some, the risk is expressed as a number which indicates rank and not an absolute value.

- Quantitative – risks are represented by the frequency or probability of a specified level of harm, from a specified activity.

Organisational arrangements for risk assessment include:

- **Plan** – what you want to achieve, who will be responsible for what, how you will achieve your aims and how you will measure your success.

- **Do** – identify your risk profile, organise your activities to deliver your plan.

- **Check** – measure your performance, assess how well the risks are being controlled and investigate the causes of accidents, incidents or near misses.

- **Act** – review your performance and take action on lessons learnt, including from audit and inspection reports.

Once hazards have been identified, the risk they pose needs to be assessed and prioritised.

Systems Failures and System Reliability

We have:

- Noted that systems can have very complicated interactions between processes and that failure of the system may need detailed investigation to discover the causes by adopting both holistic and reductionist approaches.

- Used calculations to assess the reliability of parallel, series and mixed systems.

- Considered common mode failure and the principles of HRA.

- Examined methods for improving system reliability by using reliable components, quality assurance, parallel redundancy, standby systems, minimising failures to danger, planned preventive maintenance and minimising human error.

Failure Tracing Methodologies

Principles and techniques of **failure tracing methods**:

- A **HAZOP** is:

  - A method designed for dealing with complicated systems where failures can affect other parts of the system.

  - Used, for example, at large chemical plants.

- **FTA** starts with a **Top Event** and identifies the necessary preceding events and their combination that are necessary for the Top Event. It identifies causes.

- **Event trees** start with an **Initiating Event** and look at the consequences through the failure of control measures and the effect of other events.

Exam Skills

This is an example of a 20-mark, Section B question for you to work on. It is split into two sub-questions relating to risk assessment.

You should allow 30 minutes to answer a Section B question in the exam.

QUESTION

(a) **Outline** a range of factors that should be considered in order to produce a 'suitable and sufficient' risk assessment. **(8)**

(b) Select a range of internal and external information sources **AND outline** how **EACH** source contributes to risk assessment. **(12)**

Approaching the Question

At first sight this might seem like a simple risk assessment question but you need to read it carefully.

Re-read the appropriate sections of the element and think about the terms asked about in the question. This will help to get things clear in your mind. This question is about risk assessment but not the simple five-step approach. It asks you what constitutes a 'proper' risk assessment and what information is useful to the risk assessment process.

Think about the marks available for the two parts of the question, and your timing.

Put together an answer plan for part (a) of the question around the phrase 'suitable and sufficient'. Remember that there are eight marks available for this part so the examiner is looking for all the factors required to ensure a suitable and sufficient risk assessment.

Part (b) requires a list of internal and external information sources that would be useful in the risk assessment process, but note that you also need to show how each source actually contributes to risk assessment. So it is not enough to simply provide a list of sources of information and marks would only be awarded if you made sure that you outlined the purpose of each source. Note that the question distinguishes 'internal' and 'external' information sources so this is a good way to structure your answer.

HINTS AND TIPS

As you write your answer, make sure that you refer back to the scenario described in the question to ensure that your answer has the correct emphasis.

Suggested Answer Outline

In part (a), the examiner would want you to show that you understand what is meant by the term 'suitable and sufficient' used in Regulation 3 of the Management of Health and Safety at Work Regulations 1999, so to be effective, the risk assessment should:

- Identify the significant risks arising from, or in connection with, the work.

- Contain the level of detail which is proportionate to the risk.

- Ignore insignificant risks.

- Be completed by a competent person.

- Consider all those who might be affected by the activities such as workers or others, such as members of the public.

- Make use of appropriate sources of information such as law, guidance, manufacturers' information.

- Not be expected to anticipate risks that are not reasonably foreseeable.

- Be appropriate to the nature of the work.

- Indicate the period of time for which it is likely to be valid.

In part (b), the examiner is concerned with both **internal** and **external** information sources so it makes sense to structure your answer in this way. It also helps to focus your mind because you can first concentrate on internal information sources such as accident data, etc. then move on to think about external information sources like legal standards.

Internal sources:

- Accident, damage and near-miss reports, which identify workplace hazards.

- Ill-health data, which helps to identify health hazards and the understanding of health risks.

- Trend analysis on reactive monitoring data to assist with the estimation of risk.

- Maintenance and inspection records for plant, which provide information on reliability and the identification of hazards.

- Proactive monitoring data from inspections and audits, which provides information on the reliability of controls.

- Observation of working practices, which identifies hazards and provides information on the reliability of controls.

- Training records, which provide information on training and assessment and contribute to the evaluation of risk.

- Employee knowledge of working practices and hazards, which provides information on the use and effectiveness of controls.

- Existing risk assessments, which identify hazards and provide information on risk controls.

External sources:

- Regulations, Approved Codes of Practice and HSE guidance, which identify hazards and set out risk control standards to meet legal requirements.

- National accident and ill-health data, which aid the identification of hazards and the estimation of likelihood.

- Professional and industry bodies and trade unions, who provide statistics and guidance to aid the identification of hazards, the estimation of risk and the establishment of risk control standards.

- Insurance companies, enforcement officers and consultants, who provide guidance on risk control standards.

- International bodies such as the WHO, which produces guidance on specific hazards and risk control standards.

- Manufacturers' information or data sheets, which provide hazard information.

Example of How the Question Could Be Answered

(a) The term 'suitable and sufficient' is found in Regulation 3 of the Management of Health and Safety at Work Regulations 1999. For a risk assessment to be considered 'suitable and sufficient', first of all it needs to identify the significant risks associated with the work, while containing a level of detail which is proportionate to the risk. This means that insignificant risks can be ignored. The risk assessment must be completed by a competent person who will consider all those who might be affected by the activities, i.e. not just workers but others affected, such as members of the public. The risk assessment should make use of appropriate sources of information such as law, guidance and manufacturers' information but can't be expected to anticipate risks that are not reasonably foreseeable. Finally, the risk assessment has to be appropriate to the nature of the work and also give some indication of the period of time for which it is likely to be valid and therefore when a review will be necessary.

(b) Internal and external information sources can both contribute to risk assessment.

Internal information: accident, damage, near-miss reports and ill-health data all assist with the identification of workplace hazards and the understanding of health and safety risks. Trend analysis on this type of reactive monitoring data can assist with the estimation of risk from the patterns of data collected. Maintenance and inspection records for plant will provide information on breakdowns and failures, i.e. reliability data, as well as the identification of hazards arising from plant failures. As well as reactive data, we can also use proactive monitoring data from inspections and audits to provide information on the reliability of controls. Another source of this type of information is through actually observing and checking employee working practices and the associated control measures. Training records will give information on the competency of the workforce which is also a key factor in the evaluation of risk. Finally, existing risk assessments can be useful as a starting point for determining the range of workplace hazards and the status of existing controls.

External information: Regulations, Approved Codes of Practice and HSE guidance are a primary source of information on significant hazards and the associated risk control standards required to meet legal requirements. National accident and ill-health data aid the identification of principal hazards and causes. They also provide information on numbers of incidents and prevalence of cases of ill health which is useful in estimating probabilities of occurrence. Professional and industry bodies and trade unions are able to provide statistics and guidance relating to particular types of work which again assist with the identification of hazards, the estimation of risk and the establishment of risk control standards. And insurance companies, enforcement officers and consultants are all able to give guidance on risk control standards, as can international bodies such as the World Health Organization. Finally, for particular items of plant and equipment, manufacturers and suppliers are legally bound to supply information or data sheets on their associated hazards.

Reasons for Poor Marks Achieved by Candidates in Exam

An exam candidate would achieve **poor marks** for:

- Not appreciating that 'suitable and sufficient' are specific terms used in Regulation 3 of the Management of Health and Safety at Work Regulations 1999 and therefore have a particular meaning.

- Not understanding enough about the concept of risk assessment to be able to identify basic factors such as identification of significant risks, appropriate level of detail, need for competent person and consideration of all those affected by work activities.

- Not clearly showing what the relevance of the information source is to the process of risk assessment.

- Considering a limited range of information sources such as accidents, inspections and legislation which are insufficient to warrant the 12 marks available in part (b).

Risk Control

Learning Outcomes

Once you've read this element, you'll understand how to:

1 Explain the use of common risk management strategies.

2 Outline factors to be taken into account when selecting risk controls.

3 Explain the development, main features and operation of safe systems of work and permit-to-work systems.

Contents

Common Risk Management Strategies

IN THIS SECTION...

The key risk management strategies are:

- Avoidance or elimination.

- Reduction.

- Transfer.

- Retention with/without knowledge.

The risk management aim is to develop risk controls relevant and appropriate to the risk being assessed.

Concepts Within a Health and Safety Management System

Risk management may be defined as: 'The identification, measurement and economic control of the risks which threaten the assets or earnings of a company or enterprise.'

Risk control can be split into **loss control** and **risk financing**:

- Loss control:

    - Risk avoidance.

    - Risk reduction.

- Risk financing:

    - Risk retention.

    - Risk transfer.

A strategy may consist of one or a combination of these methods.

Risk control includes loss control and risk financing

Avoidance or Elimination

- **Risk avoidance** is avoiding completely the activities giving rise to risk. For example, stop drinking alcohol at all (or never start) to avoid the risk of being arrested for being drunk and disorderly; never travel by air to avoid the risk of being involved in a mid-air collision.

- **Risk elimination** usually has a wider meaning; it implies removal of a risk **without necessarily ceasing** an activity completely, e.g. redesign of a process to remove a particular risk without stopping the activity.

Risk avoidance or risk elimination is the best solution to the problem of risk. In some cases, we will have estimated the risk of some particular operation to involve the possibility of a fatality or serious personal injury. This suggests that avoidance or elimination is an essential requirement. In eliminating one risk, you could inadvertently introduce other risks. For example, in automating a process by introducing robots to eliminate, say, the risk of manual handling, you will introduce some of the risks associated with robots. Some hazards can be avoided by completing a task in a slightly different way. For example, providing a chair for a supermarket checkout person (rather than expecting them to stand) can remove hazards associated with physical fatigue.

Reduction

DEFINITION

RISK REDUCTION

Is where risk is not avoided or eliminated entirely, but attempts are made to reduce the frequency and/ or severity of a potential loss by use of typical safety control techniques, such as engineering solutions, procedures and behavioural measures (training, etc.) to control risk at source.

Often, avoidance or elimination may not be possible or reasonably practicable or even desirable (if, for example, it would involve closing a factory with the loss of all jobs and high associated cost of redundancy). Risk reduction, while not as effective, might be a more economically viable solution.

Risk Retention

Here, the loss is to be financed from funds within the organisation, so we have to consider where the funds are to come from.

Sources of Funds

Possible sources are:

- Pay losses from current operating funds. Payments should be restricted to a maximum of about 5% of the operating costs. Losses must be predictable.

- Use an unfunded reserve, such as depreciation. This is where some large item of capital expenditure is written off over a number of years. The problem is that the fund does not actually exist except as an accounting convenience. There is no tax advantage and no actual ready cash.

- Use a funded reserve, e.g. a fund of cash or easily obtained cash. It could be a group fund. There is no tax advantage. It takes time to build up such a reserve, so care is required in the early years. There is low interest on capital. If you wish to obtain a good rate of interest, you will have to give notice before you can withdraw funds. The fund needs to gain interest, but should be readily available when required.

- Insuring through a captive insurer (see later).

- Borrowing to restore losses, which is not easy after a loss occurs. For example, if you had just had a large fire at a factory, the bank would be reluctant to lend and would make a lot of expensive conditions.

- Divert funds from planned capital investment; the company then uses funds set aside to buy an important capital item because there is a loss which has to be paid for.

If you consider each of these options, you will realise that there is no readily available, inexpensive source of finance to pay for any loss. On the other hand, there are some good reasons for considering risk retention.

Advantages of risk retention include:

- The full sum of insurance premiums is never paid out, so risk retention can be cheaper than insurance. The insurance company has to make a profit both for future finance and for its shareholders. A millionaire does not insure their car comprehensively; if it is damaged, they just buy another one. The good car driver or employer pays, through their premiums, for the poor driver or employer. Insurance is profitable for poor risk managers, but not for a good risk manager.

- Retention reduces the cost of both processing claims and the detailed accounting required. The loss occurs and you just pay out.

- If costs are allocated to departments, management becomes more risk conscious. This is a vital feature in risk management; it is pointless for a departmental manager to go all out for production profit and then have to use their profits to pay for accidents and losses.

- Losses are dealt with quickly.

You should think about each advantage and see whether it applies to your organisation.

Risk Retention – With or Without Knowledge

With knowledge means you have made a conscious decision to bear the burden of losses; **without knowledge** means it is done without any consideration whether or not to insure.

Every risk that is not transferred (to insurance) is a retained risk. Examples are:

- **Events that are insurable** – you cannot get insurance for everything. The insurance company has to be able to assess risk since they are in the business of risk management. They may quote a premium above the value you wish to insure. If you can buy a new item for the price of the premium, it is pointless to insure. Take the risk instead.

- **Losses not considered when setting up insurance** – if you do not take into account a particular possibility, you are retaining the loss. It is a case of accidental risk retention, or risk retention by default.

Many risks can be insured against

- **Hazards deliberately not insured** – you have to insure a car for third-party risks, but the choice to insure comprehensively is left to you. Risk management is all about taking a risk, where you have been able to reduce either the probability or the severity of a loss-making event.

- **Losses outside the scope of the insurance** – there are always exclusion clauses, and you do not realise their significance until you need to make a claim. The good risk manager does not find themselves in such a situation.

- **The part of the loss paid by the company (the excess)** – you can get cheaper insurance if you agree to pay the first £x of any claim.

- **The part of the loss which is above the limits of the contract** – there is often an upper limit to an insurance claim. The claimant pays if the loss exceeds that figure.

- **The person or company is unable to pay full compensation** – obtaining the cheapest insurance cover may not be sound economy if your losses put them into bankruptcy.

Risk Transfer

Transfer involves transferring the risk to another party such as by insurance – the loss is financed from funds that originate outside the organisation. The second main way is to engage a contractor who will take on the risks.

Insurance

How can you reduce insurance premiums? One way is to retain losses; another way is to accept a voluntary excess on insurance premiums and control losses.

Advantages of insurance are that the:

- Loss will be dealt with smoothly. There will be a few forms to fill in and enquiries, but the procedures are well known.

- Cash is available. The insurer can get hold of the funds quickly, though will perhaps not release them as quickly as you would like.

- Insurer can provide advice. They are dealing with this type of problem all the time and can help you to decide what is best.

Use of Specialist Contractors

Sometimes, the best way of avoiding a hazard is to make use of specialist contractors, e.g. for the removal of asbestos. In this way, the hazard is avoided by employees and the task is carried out professionally and in compliance with current legislation. A reputable company with suitably trained personnel and a good safety record should be used.

Risk Sharing

Risk management is really a type of risk sharing and involves financing risks that are manageable and transferring those that are not.

Methods include:

- A deductible portion of excess – you pay the first part of each claim.

- Re-insurance.

- Co-insurance – the insurer pays a percentage of the claim. This is another way of reducing a premium. You share the risk with the insurer by paying not only an excess but a percentage of the losses which fall within a certain price range, paying another percentage of those in another range, and the insurer paying all losses above a set figure.

An Important Point

A good risk manager will:

- Not insure where they have eliminated a risk.

- Consider very carefully those areas where they have significantly reduced the risk.

- Pay for the retained risks where it is cheaper than insuring.

However, if they ask for insurance for just some risk, the insurer is going to be wary. They are only using insurance as a last resort, so there must be a problem they cannot solve.

A good risk manager will make savings in the area of insurance

Circumstances When Each of the Previous Strategies Would Be Appropriate

Risk Avoidance

If risks can't be managed to an acceptable level, it might be necessary to not proceed with the activity. The choice of avoiding an activity is limited in certain business sectors where there could be regulatory obligations to provide certain services. If the costs of achieving a project are too high due to necessary risk control then the only response might be to end the project. However, risk avoidance as a business option should not be confused with risk aversion where sensible risk management has not been applied and the risks have been incorrectly assessed.

Risk Retention

The risk is retained in the organisation where any consequent loss is financed by the company. **Risk retention with knowledge** means that there is no further action planned to deal with it, perhaps because there are no control options available or because the only options are unacceptable or cannot be implemented yet. Risk retention is a conscious decision based on the findings of the risk evaluation process but should be kept under review in case circumstances change.

Risk retention without knowledge because of a failure to identify or appropriately manage risks is not a viable option since it is a consequence of an ineffective risk assessment system.

Risk Transfer

This refers to the legal assignment of the costs of certain potential losses from one party to another, the most common way being by insurance. It is also possible to transfer the whole risk to another organisation but there is still the possibility that the organisation to which the risk is transferred might not manage the risk effectively. In practice, risk transfer tends to be used in conjunction with one or more of the other risk management options.

Risk Reduction

Here, the risks are systematically reduced through control measures, according to the hierarchy of risk control (see later). This is the most common way to manage risks and aims to reduce the likelihood and/or severity of undesired consequences through preventive measures and/or contingency plans.

Factors in the Selection of an Optimum Solution Based on Relevant Risk Data

The selection of the optimum solution must take into account the type of organisation and the relevant risk data. The risk assessment will be a vital part of the exercise. If the probability is high and the severity is also high, then it will be important to do a great deal and spend a lot of finance to achieve a valid solution. If the probability or the severity is low, then it will not warrant too great an expenditure.

Study the following table carefully:

Probability	Severity	Action
Definite	High	Eliminate
	Medium	Fund (cheaper than insurance)
	Low	No action -- operating expense
High	High	Eliminate or reduce probability or severity
	Medium	Reduce severity
	Low	Retain as an operating expense
Medium	High	Reduce severity
	Medium	Reduce severity or transfer
	Low	Retain as an operating expense
Low	High	Fund or insure
	Medium	Fund
	Low	Retain as an operating expense
Remote	Catastrophic	Insure, or fold company
	High	Fund, insure, or fold company
	Medium	Fund or retain as an expense
	Low	No action

Other important factors include:

- **Present State of Technology**

    With technology evolving all the time, new solutions become available; as computers have improved, the price of this technology has reduced. The health and safety practitioner must keep up-to-date with technology and consider how it can contribute to safety solutions.

- **Public Expectancy**

    "After a disaster, a journalist often asks: 'Can you guarantee that nothing like this will ever happen again?' Remember that human beings make mistakes, and no machine is infallible. Earthquakes occur without warning, and we can do little to control the effects of freak weather.

    A major problem is that the general public is never realistic in assessing relative risk. Car accidents cause more deaths than public transport, but seldom hit the national headlines. Atomic energy is probably one of the safest power sources. Most deaths are from cancer, heart disease and stroke, and very few are from industrial causes. Most people would make a poor job of rating industries by their accident potential.

'Can you guarantee that nothing like this will ever happen again?'

- **Legal Requirements**

  'Learning by accident' is an apt way of describing how safety legislation has been developed over the years. Industrial accidents and disasters are the basic reason for much of British legislation. Mines and factories were the cause of fatalities, so legislation was enacted to control them. Now legislation is more general. This sets standards that have to be adhered to; these are the basis of minimum standards. The selection of controls will need to consider both criminal law and civil law. Regulations made under the **Health and Safety at Work, etc. Act 1974** set out minimum standards for control of machinery, chemicals, electricity, radiation, etc. There may also be contractual obligations that have to be met and will determine standards of control. The need for employer's liability insurance may require insurance companies to dictate the standards of health and safety in the workplace that need to be achieved before insurance cover is provided.

- **Economic State of the Company**

  The economic state of the company is no excuse for not meeting legal standards – it can be used as a reason for not going for a higher standard. It is not good economic sense to skimp on safety, since all accidents produce a loss. However, a company with vast profits can afford to spend more than one with financial restraints. Companies' economic goals will influence the approach to risk control. These may range from simple cost-covering to survive, to profit maximisation. Risk management must balance the cost of controls against the estimated reduction in potential loss from risks.

- **Levels of Insurance Premiums**

  Premiums are set by the level of claims. The insurance company is in business to make a profit. Good companies pay their premiums and do not make claims; it is the bad companies who benefit from insurance. If the premiums are reasonable then it is better to take the precaution of insurance. However, if premiums become excessive then it will be better to retain the risk. One person may insure their car comprehensively, because they can get a cheap rate and a 70% no-claim discount. Another person may settle for third party, fire and theft because the premium for comprehensive is almost the same as the value of the car.

- **Confidence of the Company in the Benefits of Risk Management and in the Competence of the Risk Manager**

  A good company, with good control of risk, will opt to retain risk rather than insure or transfer the risk. As the situation and the confidence improve, there will be increased movement toward this method of solution. If the company is able to use a captive insurance company, there will be less reliance on outside risk transfer.

- **Human Factors**

  Accidents and incidents have an associated direct cost but can also influence the culture of the organisation. Frequent loss-making events can have a bad effect on morale, which can lead to reductions in efficiency and higher overall costs. Consequently, a wish to improve industrial relationships can influence the approach to risk control measures.

Principles and Benefits of Risk Management in a Global Context

All activities involve risk, therefore organisations have to manage risk by identifying it, analysing it and evaluating whether the risk needs to be controlled in order to satisfy their risk criteria.

TOPIC FOCUS

BS ISO 31000:2018 Risk Management – Guidelines sets out the following principles of effective risk management:

(a) Risk management is an integral part of all organisational activities.

(b) A structured and comprehensive approach to risk management contributes to consistent and comparable results.

(c) The risk management framework and process are customised and proportionate to the organisation's external and internal context related to its objectives.

(d) Appropriate and timely involvement of stakeholders enables their knowledge, views and perceptions to be considered. This results in improved awareness and informed risk management.

(e) Risks can emerge, change or disappear as an organisation's external and internal context changes. Risk management anticipates, detects, acknowledges and responds to those changes and events in an appropriate and timely manner.

(f) The inputs to risk management are based on historical and current information, as well as on future expectations. Risk management explicitly takes into account any limitations and uncertainties associated with such information and expectations. Information should be timely, clear and available to relevant stakeholders.

(g) Human behaviour and culture significantly influence all aspects of risk management at each level and stage.

(h) Risk management is continually improved through learning and experience.

The **benefits** of risk management to an organisation are self-evident in terms of loss prevention and business disruption, but there are a number of specific benefits that can be used to support it at an organisational level:

- Increased likelihood of achieving objectives.

- Encouragement of proactive management.

- Improved identification of opportunities and threats.

- Compliance with relevant regulatory requirements.

- Improved governance, stakeholder confidence and trust.

- A reliable basis for decision-making and planning.

- Improved risk control, loss prevention and incident management.

- Effective allocation and use of resources for risk control.

- Improved operational effectiveness and efficiency.

- Enhanced health and safety performance.

Enhanced health and safety performance is just one of the benefits of risk management

Organisations with multiple locations worldwide still need to consider how they intend to manage risk and the associated benefits but face the challenge of how to manage safety across their global sites where standards and local conditions may vary considerably.

Large multinational organisations may decide to consistently apply the principles and seek the benefits examined earlier by adopting the same Safety, Health and Environmental (SHE) management standards for all of their locations globally. One benefit of this approach is that it is easier to audit (one company, one standard) and allows comparisons to be made across global sites.

Other organisations may decide simply to only comply with the local SHE standards. This is likely to be a more cost-effective approach, and also legally compliant, but the standard to be achieved may be lower.

Link Between the Outcomes of Risk Assessments and the Development of Risk Controls

The aim of the risk management process is to develop risk controls relevant to the risks being assessed. The outcome of the risk assessment process follows from risk evaluation and involves making decisions about risk controls and the priority for their implementation. Risk evaluation involves comparing the level of risk found during risk analysis (which considers the causes and sources of risk, their consequences and the likelihood of occurrence) with established risk criteria. This comparison then informs the decision about the need for controls.

The risk assessment process may conclude the need to implement additional controls, change controls or continue with no controls at all.

We have seen already that controls may be implemented to:

- Avoid risk.
- Retain risk.
- Transfer risk.
- Reduce risk.

In order to decide which controls to implement, an assessment is needed based on the risk criteria:

- The relative cost benefits of the control options considered.
- Any regulatory requirements or social responsibility factors that might override a cost-benefit analysis.
- Any additional risks that might be introduced by a particular control.

The development of risk control options therefore follows directly from the risk assessment process.

STUDY QUESTION

1. What are the main risk management strategies?

(Suggested Answer is at the end.)

Factors to Be Taken Into Account When Selecting Risk Controls

IN THIS SECTION...

- The **Management of Health and Safety at Work Regulations 1999** require control measures to be considered using the hierarchy:

    - Avoid risks.

    - Evaluate.

    - Combat at source.

    - Adapt work to the individual – ergonomics.

    - Adapt to technical progress.

- Control measures may be classified as being technical, procedural or behavioural.

- The choice of control measures adopted should take account of:

    - Use in the long or short term.

    - Applicability.

    - Practicability.

    - Cost.

    - Proportionality.

    - Effectiveness.

    - Legal requirements.

    - Competence/training needs.

General Principles of Prevention

Having identified the risks, measured their effect on the company and developed some kind of priority, we then have to do something about them. In practice, we are probably doing each of these stages at the same time:

- Recognise.

- Measure.

- Evaluate.

- Control.

- Monitor.

- Review.

Principles of prevention

TOPIC FOCUS

Preventive and Protective Measures

Schedule 1 to the **Management of Health and Safety at Work Regulations 1999** sets out the principles of prevention referred to in **Regulation 4**. The principles of prevention provide a framework against which to make judgments on the preventive and protective measures which are required to control risks to an adequate level.

The principles are:

- **Avoiding Risks**

  Not using the material (e.g. toxic chemicals) or not carrying out the activity (e.g. excavations) eliminates the need for control.

- **Evaluating the Risks Which Cannot Be Avoided**

  For these risks, a suitable and sufficient risk assessment must be carried out.

- **Combating the Risks at Source**

  Control the risk as close to the point of generation as possible to prevent its escape into the workplace (e.g. extract dust directly from a circular saw blade using Local Exhaust Ventilation (LEV)).

- **Adapting the Work to the Individual**

  Consider, in particular, the design of workplaces, the choice of work equipment and the choice of working and production methods, with a view to alleviating monotonous work and work at a predetermined work-rate and to reducing their effect on health.

  The traditional approach has always been for the person to adapt to the machine or process. This measure requires the employer to carefully consider ergonomic principles and design the work to suit the person.

- **Adapting to Technical Progress**

  Many risks disappear from the workplace as better processes and methods are introduced. For example, the replacement of traditional machine tools by Computer Numerical Control (CNC) machines, primarily for production efficiency, also removes the need for manually adjusted guards on lathes and milling machines.

- **Replacing the Dangerous with the Non-Dangerous or the Less Dangerous**

  This is always a key aim, and an example of this is the replacement of the metal-cased, hand-held mains electric drill by rechargeable, battery-operated, plastic-cased drills.

- **Developing a Coherent Overall Prevention Policy**

  Such a policy should cover technology, organisation of work, working conditions, social relationships and the influence of factors relating to the working environment.

  This embodies the principles of risk management and requires the employer to look at all aspects of the health and safety management system rather than simply concentrating on basic workplace precautions.

- **Giving Collective Protective Measures Priority over Individual Protective Measures**

  A safe place of work should be the main priority rather than a safe person, so control of noise at source should be the aim rather than the issue of hearing protection.

- **Giving Appropriate Instructions to Employees**

Determine Technical/Procedural/Behavioural Control Measures

Categories of Control Measures

Control measures are often categorised into one of three different types:

- **Technical** – the hazard is controlled or eliminated by designing a new machine or process, or by producing some guarding measure.

- **Procedural** – such as a safe method of work, e.g. introducing permit-to-work systems as part of a safe system of work.

- **Behavioural** – involves education and training of operatives, putting up notices and signs, using protective equipment and generally making employees aware of the risks – changing the 'safety culture' of the organisation.

General Hierarchy of Control Measures

In dealing with risks, we must establish an order of treatment. A quick search on the internet will make you realise that there are a number of different hierarchies, many of which are very similar; some are specific to control of chemicals or machinery guarding.

ISO 45001 (*Occupational health and safety management systems – Requirements with guidance for use*) specifies the following hierarchy which should be considered when determining controls, or considering changes to existing controls, in order to reduce risks:

- Elimination.

- Substitution.

- Engineering controls.

- Signage/warnings and/or administrative controls.

- Personal Protective Equipment (PPE).

TOPIC FOCUS

Hierarchy of Control Measures

From ISO 45001:

- **Elimination (Technical)**

  Stop using the process, substance or equipment, or use it in a different form.

- **Substitution (Technical/Procedural)**

  Replace a toxic chemical with one that is not dangerous or less dangerous. Use less noisy pumps.

- **Engineered Controls (Technical/Behavioural)**

  Re-design the process or equipment to eliminate the release of the hazard so that everyone is protected; enclose or isolate the process or use of equipment to capture the hazard at source and release it to a safe place, or dilute to minimise concentration of the hazard, e.g. acoustic enclosures, use of LEV.

- **Signage/Warnings and/or Administrative Controls (Procedural/Behavioural)**

  Design work procedures and work systems to limit exposure, e.g. limit work periods in hot environments, develop good housekeeping procedures. Controls may also include: use of signs, training in specific work methods, and supervision.

- **PPE (Technical/Behavioural)**

  Respiratory protective equipment, gloves, etc. – only protects the individual.

The items at the top of the list are often long-term objectives and are the responsibility of management. They are the most effective, but more costly to implement. The items toward the bottom of the list can be short term and quickly put into place, but are the least effective. It may be impossible or prohibitively expensive to eliminate a hazard in a practical situation. On the other hand, you will get very few marks in the examination if your solution to a practical situation is to issue a pair of gloves, or just suggest that an employee takes more care.

Another example of a hierarchy that is sometimes quoted is:

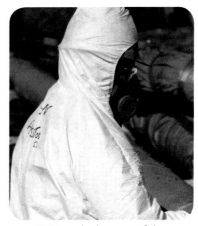

PPE is at the bottom of the hierarchy of controls

- Total elimination or avoidance of the risk at its source.

- Reduction of the risk at its source.

- Contain the risk by enclosure of some kind.

- Remove the employee from the risk.

- Reduce the employee's exposure to the risk.

- Use PPE.

- Train the employee in safe techniques.

- Make safety rules, or issue instructions.

- Tell the employee to be careful.

Factors Affecting Choice of Control Measures

Long Term/Short Term, Applicability and Costs

Those points which appear earlier in the list of control measures will be the most effective in reducing the risk, but are usually the more expensive and take much longer to put in place, so can be viewed as **long-term objectives**. Although, in practice, it might be technically possible to achieve total elimination of a hazard, the costs involved and the benefits achieved may mean that it does not pass the test of 'reasonably practicable'.

Many of the improvements in safety standards have been due to the reduction in the workforce and increasing mechanisation. Computers can be used to control many operations and eliminate the use of people in risky situations. However, they cannot think, and sometimes the choice is not between right and wrong (1 and 0 to a computer) but between the lesser of two wrongs.

The methods shown lower in the list of control measures are usually the cheaper options. They can be put into operation quickly, and give some measure of risk reduction, but their effect is of short duration. PPE, although near the bottom of the hierarchy, may be acceptable for non-frequent exposure, such as in maintenance tasks.

Proportionality

It is the responsibility of organisations to take ownership of their risks and therefore to take proportionate (sensible) steps to manage those risks. This means focusing attention on the significant risks that cause injury and ill health, not the trivia or everyday low risks. Proportionality is achieved by concentrating on the real risks – those that are reasonably likely to cause a significant level of harm – and not wasting valuable time and resources on unlikely events with low-level outcomes.

Effectiveness of Controls

No one control measure can be 100% effective, so when evaluating which measure to adopt you have to take into account its effectiveness. PPE is of limited benefit because it only protects the person wearing it and not necessarily all those at risk; it may be uncomfortable or inconvenient to wear. The more effective the control, the greater consideration should be given to its use.

Legal Requirements and Standards

In some circumstances, legislation dictates the standard required. For example, if there is risk of rollover from mobile work equipment such as a forklift truck, then the **Provision and Use of Work Equipment Regulations 1998** require a structure that ensures that the work equipment does no more than fall on its side.

Competence of Personnel and Training Needs

Clearly, the control measures adopted for a specific situation must be such that the user is competent to use them without them creating a risk to the user or others. This may mean additional training and supervision which are an added cost.

STUDY QUESTIONS

2. Outline the three broad categories of control measure.

3. List the factors that should be considered when choosing control measures.

(Suggested Answers are at the end.)

Safe Systems of Work and Permit-to-Work Systems

IN THIS SECTION...

- The following need to be considered when devising a safe system of work:
    - People.
    - Equipment.
    - Materials.
    - Environment.
- A permit to work is a formal, written document of authority to undertake a specific procedure and is designed to protect personnel working in especially hazardous areas or activities. It details the:
    - Task.
    - Significant hazards.
    - Control measures to be used before work starts and during work.
    - Checks to establish normal work can resume.
    - Persons authorised to undertake the task and those responsible for monitoring it.
- A safe system of work is developed from a risk assessment of the task.

Safe Systems of Work

Legal Requirements: Health and Safety at Work, etc. Act 1974

> **DEFINITION**
>
> **SAFE SYSTEM OF WORK**
>
> One where the work is organised in a logical and methodical manner so as to remove the hazards or minimise the risks.
>
> The term is not defined in legislation but the HSE have given a useful working definition:
>
> *"A safe system of work is a formal procedure which results from systematic examination of a task in order to identify all the hazards. It defines safe methods to ensure that hazards are eliminated or risks minimised."*
>
> Source: IND(G)76L *Safe systems of work*, HSE

As you know, this is the most important piece of legislation as far as health and safety is concerned. With regard to safe systems of work, we will be concerned mainly with **Section 2** of the Act:

> *"It shall be the duty of every employer to ensure, so far as is reasonably practicable, the health, safety and welfare at work of all his employees."*

This general duty is all-inclusive. The next subsection starts with the phrase: *"Without prejudice to the generality of an employer's duty under the preceding subsection, the matters to which that duty extends include in particular..."*.

This means that a number of specific requirements are going to be mentioned. However, if the employer complies with all of these requirements, they may still need to do more in order to comply with the requirement of this subsection.

The next subsection states that the employer has to provide and maintain:

"*plant and **systems of work** that are, so far as is reasonably practicable, **safe** and without risks to health.*"

In several instances, a duty is qualified by the phrase, "*so far as is **reasonably practicable***". This phrase has a precise legal definition. **Reasonably practicable** implies that a computation must be made in which the quantum of risk is placed in one scale and the sacrifice involved in the measures necessary for averting the risk (whether in time, money or trouble) is placed in the other, and that if it can be shown that there is a gross disproportion between them – the risk being insignificant in relation to the sacrifice – the employers discharge the onus upon them.

This means that the employer can only use costs as the reason for not doing something when the risk is insignificant when compared with the cost of eliminating the risk. The burden of proof lies with the employer. The size or financial position of the employer is not taken into account in this calculation.

Practical Requirements

In developing a practical safe working system, adequate provision must be made for:

Mandatory signs to promote a safe working environment

- Safe design of plant and equipment.
- Safe installation of plant and machinery.
- Safe maintenance of plant, equipment and premises.
- Safe use of plant, equipment and tools through proper training and supervision.
- Documented, planned maintenance procedures.
- Safe working environment (ventilation, heat and light).
- Adequate and competent supervision.
- Trained and competent employees.
- Proper enforcement of the safety policy and associated rules.
- Adequate personal protection for vulnerable employees.
- Documented procedures for the issue of protective clothing.
- Dissemination of health and safety information to all the employees.
- Regular reviews (not less than once a year) of all job safety instructions and methods of work to ensure that:
    - There is continued compliance with legislation.
    - Plant modifications are taken into account.
    - Substituted materials are taken into account.
    - New work methods are incorporated.
    - Systems still work safely in practice.
    - Advances in new technology are exploited.
    - Safety precautions are adjusted to take into account accident experience.

Components of a Safe System of Work

A safe system of work constitutes the bringing together of:

- People.
- Equipment.
- Materials.
- Environment.

Systems of work must have a logical, well thought-out approach when compared with **methods of working**, which often merely evolve over time. In a system of work, there is a logical progression, from hazard identification and prediction, so that such hazards are eliminated or controlled.

Safe systems of work should fully identify and document all the hazards, safety precautions and safe working practices associated with all activities performed by employees. The analysis should be capable of identifying any unsafe work methods. There must also be a system of monitoring safety performance and for publishing information about such performance.

Job safety training is also an integral part of the safe working system; there must be a basic commitment to provide high standards of safety training for all operatives, new entrants, line managers and the safety practitioner and safety representatives.

When is a Safe System Required?

Many hazards are clearly recognisable and can be overcome by physically separating people from them, e.g. by using guarding on machinery.

A safe system of work is needed when hazards cannot be physically eliminated and some element of risk remains. You should apply these principles to routine work as well as to more special cases, such as:

- Cleaning and maintenance operations.
- Making changes to work layouts, materials used or working methods.
- Employees working away from base or working alone.
- Breakdowns or emergencies.
- Controlling activities of contractors on your premises.
- Loading, unloading and movement of vehicles.

Five Steps to Developing a Safe System

There are five basic steps to developing safe systems of work:

Step 1: Assessing Work Tasks

All aspects of a particular task/job or operation must be looked at and put in writing to make sure nothing is overlooked. The safe system of work should be based on a thorough assessment of the task the system is to cover. Assessment should be done by supervisory staff together with the workers involved, so any assumptions supervisors might have about methods of work do not differ from reality. Workers may also be in the best position to help with preparing a safe system of work. Consulting workers exposed to risks either directly or indirectly is also a legal requirement.

Account must be taken of:

- What is used: plant and equipment, substances, potential machinery failures, the task's electrical needs.
- Error sources: possible human failures, shortcuts.
- Where the task is carried out: the working environment and its protection needs.
- How the task is carried out: procedures, potential work method failures, task frequency, training needs.

Step 2: Identifying Hazards and Assessing Risks

List the task's elements and clearly identify the associated hazards using a risk assessment. Note that the law requires a 'suitable and sufficient' assessment of all risks that employees and others may be exposed to.

The way the analysis is done depends on the nature of the task/job or operation. If what is being considered involves high loss potential then formal hazard analysis techniques such as a Hazard and Operability Study (HAZOP), Fault Tree Analysis (FTA) or failure modes and effects analysis should be considered.

But if the potential loss is lower, a simpler approach such as job safety analysis may be used. If the employer can't remove hazards and risks remain, then procedures to ensure a safe method of work must be worked out.

Step 3: Defining Safe Methods

Instructions must be given by supervisors or managers. Leaving workers to come up with their own method of work isn't a safe system of work!

The chosen method can be explained verbally and/or in writing. Jobs require different types of safe systems – depending on the levels of risk:

- A very low-risk job may require only following safety rules or a previously agreed guide (which may or may not be in writing).
- A very high-risk job may require an all singing, all dancing formal written permit to work system.

Risk level/safe system type:

- Very high: permit to work.
- High: written safe system or permit.
- Moderate: written safe system.
- Low: written safe system.
- Very low: verbal instruction (with written backup such as safety rules).

Documented methods should as far as possible be written in a clear, non-technical style. Easy-to-read simple summary sheets containing all the key points should be used. Whatever the method, the process is likely to involve:

- setting up the task along with any necessary authorisation,
- planning job steps,
- stating the approved safe working methods including, if appropriate, how to get to and from the task area,
- conditions which must be confirmed before work starts, and
- dismantling/disposing of equipment or waste at the end of the task.

Step 4: Implementing the Safe System

Employers have a duty to come up with and maintain safer systems of work. They must make sure their safe systems are effectively put into operation. They must be reviewed and revised to take account of changed conditions or accidents. Training is required for all concerned.

Employees must be:

- Adequately trained in specific systems of work.

- Competent to carry out the work safely.

- Aware of the systems and hazards which the safe methods aim to remove/reduce.

It is vital that everyone appreciates the need for the system and its role in preventing accidents.

Everyone should appreciate a safe system of work

Particular training might therefore include:

- Why the safe system is needed.

- What is involved in the work.

- The identified hazards.

- The precautions that have been decided.

For the system to be successful there has to be adequate communication. The system's details should be fully understood by everyone who works with it.

The importance of discussing the proposed system with those who will have to work under it and those who have to supervise its operation cannot be underlined enough. Safe systems require work to stop when a problem appears that is not covered by the system.

Work should not re-start until a safe solution is found.

Step 5: Monitoring the System

The safe system of work must be carried out on each and every occasion. Effective monitoring means regularly checking to make sure the system is still appropriate for the task – and that it is being fully complied with.

It is no good only checking after an accident!

Simple questions may include:

- Do workers continue to find the system workable?

- Are laid-down procedures being carried out?

- Are procedures still effective?

- Have there been any changes that require the system to be revised?

A devised system that's not followed is **not** a safe system of work! Find and fix problems – there must always be sufficient supervision if the system is to going to be followed and work carried out safely.

The level of required supervision depends on the particular employee's experience and the complexity and risks of the task. Safe systems of work are associates of, not replacements for, stronger prevention controls such as good equipment design or physical safeguards.

Permit-to-Work Systems

Need for Permit-to-Work Systems

DEFINITION

PERMIT TO WORK

A formal written document of authority to undertake a specific procedure and is designed to protect personnel working in hazardous areas or activities.

Permits are used for high-risk activities that require complex precautions and clear communication of additional controls. They are also frequently used where operations are non-routine, or if new hazards are introduced and require communication.

In many cases, it is impossible or extremely unrealistic to eliminate a risk totally. Even when the risk has been reduced, we are left with no alternative but to train someone in the skill to recognise the risks involved, and the knowledge of how to minimise them; and then, in effect, say: 'Be careful'.

A procedural way of taking every precaution possible is to institute a permit-to-work system. This is a management system that is supported by, and makes use of, permits to work, which are formal documents specifying the work to be done, hazards, and the precautions to be taken. The permit provides a clear written record, signed by a responsible manager or supervisor, that all foreseeable hazards have been considered and all the necessary actions have been taken. It **must** be in the possession of the person in charge of the work before work can commence.

TOPIC FOCUS

A permit to work has four main sections:

- **Issue** – the controls that must be implemented for the work to take place are defined. The permit is issued by an authorised person.

- **Receipt** – the workers sign onto the permit to signal that they accept the conditions of the permit and understand the hazards and control measures detailed in it.

- **Clearance/return to service** – once the work is complete, the workers sign to say they have finished and are leaving the job site to allow normal work to resume, usually also indicating whether the work is complete or not.

- **Cancellation** – control of the work area is accepted back by the issuer and the permit is cancelled. Extensions may be granted if additional time is required to complete the work.

HINTS AND TIPS

Do not mistake a 'permit-to-work system' for a 'safe system of work'. Rather, a safe system of work may require a permit-to-work system to be adopted as part of its overall systematic control of risk. The safe system of work should in itself be considered as part of the quality control procedures of an organisation.

Permit-to-work systems are only as good as those who design, implement and monitor them and so may be relatively easy to defeat.

When designing a permit-to-work system, it is essential that the company culture is established to support the system. For a permit system to be worthwhile, there needs to be an appropriate level of detail within the system (not paying 'lip service' but at the same time not being overly complex or bureaucratic), as well as a commitment to train permit-issuers and ensure that only authorised persons issue the permit. It must also be accepted that a good permit will take a little time to issue. Failure to correctly adhere to the permit-to-work system was a contributing factor in the Piper Alpha disaster.

Let us look at a permit-to-work system as an **example of a systematic means of controlling risk**.

Whenever maintenance or other temporary work of a potentially hazardous nature is to be carried out within the plant, some sort of permit-to-work system is essential.

Jobs likely to require a permit-to-work system include:

- Working in confined spaces.
- Hot work on plant containing flammable dusts, liquids, gases or their residues.
- Cutting into pipework containing hazardous substances.
- Work on electrical equipment.

Most accidents can be attributed in one way or another to human error. To achieve a high degree of safety, we have to eliminate human error as much as possible by using a system which requires formal action. Permit-to-work systems try to ensure that formal action is taken by providing a written and signed statement to the effect that all the necessary actions have been taken. The permit **must** be in the possession of the person in charge of the operation before work begins.

Cutting with an oxyacetylene torch requires a permit to work

Ideally, the control of permit-to-work systems throughout the company should be the overall responsibility of one person. That person should appreciate the existence of hazards and know how to eliminate them. They must have the necessary authority to instruct responsible people in the organisation to make safety recommendations on matters requiring their specialist knowledge.

The person responsible must also have authority to co-ordinate the efforts of everyone concerned with the provision of safe working conditions. If a permit is issued, they must make sure that everyone involved understands the terms of the permit and follows its instructions down to the smallest detail.

These provisions should be extended to any outside contractors taking part and it must be made clear that their employees must not, in any circumstances, begin work until the safety precautions and procedures have been fully explained to them.

Essential Features of a Permit-to-Work System

MORE...

The HSE publication, HSG250 *Guidance on permit-to-work systems – A guide for the petroleum, chemical and allied industries*, contains useful guidance in this area and is recommended reading.

You can access it at:

www.hse.gov.uk/pubns/priced/hsg250.pdf

Permits should:

- Define the work to be done.
- Say how to make the work area safe.
- Identify any remaining hazards and the precautions to be taken.
- Describe checks to be carried out before normal work can be resumed.
- Name the person responsible for controlling the job.

Initial Steps

Before work begins, the following general safety precautions should be observed where applicable:

- Electrical or mechanical isolation of the plant.
- Isolation of the machine or equipment area.
- Locking or blanking off of water, steam, acid, gas, solvent, and compressed air supplies.
- Erection of scaffolding.
- Provision of temporary guards (or other like equipment) to make the job safe.

The permit to work should include the following basic elements:

- Permit title.
- Permit number.
- Reference to other relevant permits or isolation certificates.
- Job location.
- Plant identification.
- Description of work to be done and its limitations.
- Hazard identification – including residual hazards and hazards introduced by the work.
- Precautions necessary. Person(s) who carries out precautions, e.g. isolations, should sign that precautions have been taken.
- Protective equipment.
- Authorisation. Signature confirming that isolations have been made, and precautions taken, except those which can only be taken during the work. Date and time duration of the permit.
- Acceptance. Signature confirming understanding of work to be done, hazards involved and precautions required. Also confirming permit information has been explained to all workers involved.

- Extension/shift hand-over procedures. Signatures confirming checks made that plant remains safe to be worked upon, and new acceptor/workers made fully aware of hazards and precautions. New time expiry given.

- Hand-back. Signed by acceptor certifying work completed. Signed by issuer certifying work completed and plant ready for testing and re-commissioning.

- Cancellation. Certifying work tested and plant satisfactorily recommissioned.

A Formal Document

The permit to work is always based on a formal document, the format and details of which will vary according to circumstances (see following example).

In addition to the safeguards outlined above, the specific safety precautions to be taken should be itemised.

The document should be valid only for a limited period, depending on the nature of the work and associated hazards.

HEPWORTH BUILDING PRODUCTS LIMITED
PERMIT TO WORK

1.   ISSUE

To _____

In the employ of _____

For the following work to be carried out:

_____
_____
_____
_____
_____

I hereby declare that it is safe to work on the following apparatus and that the safety measures detailed below have been carried out.

Here state apparatus on which it is safe to work _____
_____
_____

ALL OTHER PARTS ARE DANGEROUS

Here state exactly at what points isolating steam, water, air, gas valves, or radioactive shutters have been shut and locked off and what motors have been locked off and isolated.

_____
_____
_____

Signed _____
Being the Senior Authorised Person

Time _____
Date _____

Note: After being signed for the work to proceed, the receipt must be signed by, and The Permit retained by the person in charge of the work until the work is suspended or completed.

2.   RECEIPT

I hereby declare that I understand that the plant specified on this permit is safe to work upon and that this Permit applies to this plant only.

Signed_____
being the person in charge of the work on the apparatus upon which it is safe to work

Time _____
Date _____

Note: The apparatus mentioned must not be recommissioned until this clearance has been signed and The Permit returned by the person in charge of the work and cancelled.

3.   CLEARANCE

I hereby declare that all men under my charge have been withdrawn and warned that it is no longer safe to work on the apparatus specified in this permit and that all gear and tools are clear and that all guards have been replaced.

Signed _____Being the person in charge of the work

Time _____Date _____

4.   CANCELLATION

I hereby declare that this Permit and all copies of it are cancelled.

Signed _____Being the Senior Authorised Person

Time _____Date _____

An example of a permit to work form

All the methods to be used and precautions to be taken should be:

- Carefully discussed and agreed beforehand.

- Clearly stated on the permit.

The number of permits issued should be kept to the minimum conducive to the efficient manning of the plant.

Basic Principles of Operation

TOPIC FOCUS

Certain principles need to be observed for the operation of an effective permit-to-work system.

- **Hazard Evaluation**

  This involves recognising types of hazards which may be encountered, and then devising the means of eliminating or overcoming them. The best way of achieving this in the long term is by introducing a hazard appraisal programme, which can be used to formulate a long-lasting system of precautions. A major problem may be that work is often performed under emergency conditions and there is little time available for a detailed appraisal to be made.

- **Precaution Planning**

  All planning associated with the permit must be carried out by a competent person who should have sufficient, detailed knowledge of the hazards of the process or plant that they can formulate the plan properly. The person must have the necessary position of authority for their instructions to be recognised and complied with. They should also have an adequate knowledge of the legal requirements, and of technical terms such as 'isolate', 'lock off', and 'blank off', as they apply to the permit-to-work system.

- **Instructing the Supervisors**

  Those people responsible for the work should be carefully briefed by the person issuing the permit. The instructions in the permit must be fully understood, and this is best achieved by direct questions and answers to supplement the written word.

- **Issuing the Permit**

  The permit, which should be completed and signed by the issuer, must be given to the person in charge of the work (who signs for it). Sufficient copies must also be given to plant or site management and supervisory staff who may be involved, especially where they need to be kept informed of work progress. An additional copy of the permit should be exhibited nearby during the time it remains in force.

- **Monitoring the Permit**

  A permit-to-work system is only as good as the people who design, implement and monitor its application and so can be easy to defeat. Monitoring is therefore a crucial element. The initial explosion and resulting fire in the Piper Alpha disaster of 1988 was caused by poor adherence to a permit-to-work system. Regular monitoring is essential to ensure:

  - Full and accurate completion of the documentation, including signing off.

  - That the safe work practices specified in the permit are complied with.

General Application

Here are some examples of the types of hazardous situations in which permit-to-work systems should be used.

Electrical Equipment and Supplies

There is considerable evidence that the hazards associated with electricity are either not understood or are treated in too casual a manner. Because of the high level of risk involved and the serious consequence of switching errors and other careless mistakes, it is essential that a comprehensive safety system is put into operation whenever work is to be started on high-voltage equipment.

Any work on substation equipment **must** be covered by a permit-to-work system if safe working conditions are to be ensured and any electrical work should only be carried out by a qualified electrician.

Machinery

The biggest risk to maintenance workers is that they may be injured if machinery is started up while work is in progress. This is often because the men carrying out the maintenance work are hidden from the sight of persons at the plant controls. The machines may be set in motion as a result of some misunderstanding, negligence, or lack of knowledge and, unless the motive power is isolated and cannot be reconnected without specific authority, an accident might easily occur.

Overhead Travelling Cranes

If a person is likely to be struck by a crane while working on or near the crane track, effective measures should be taken to ensure that the crane cannot approach within six metres of the working place. These measures should include a provision to warn anyone working above floor level of the approach of the crane. It is virtually impossible to ensure full compliance with these requirements unless a permit-to-work system is adopted. In such circumstances, it is the only real safeguard.

Chemical Plant

Under normal operating conditions, chemical plant is designed to work safely. During maintenance, repair or sometimes commissioning conditions, however, hazards may be introduced or work may have to be carried out which could expose the workers to danger unless carefully planned safety procedures are adopted.

Provision of warnings should be in place for travelling cranes

Each job would have to be considered individually, because the hazards likely to be encountered (involving flammable, toxic and corrosive liquids or gases, and explosive atmospheres) vary considerably.

Formulating a permit-to-work system in a chemical plant demands a wide technical knowledge and a high degree of authority on the part of the person issuing the permit. It is likely, therefore, that only a few specialist managers will possess the necessary competence.

Radiation Hazards

The **Ionising Radiations Regulations 2017** may require that a permit-to-work system is operated to prevent the ingestion, inhalation, or other absorption of radioactive material into the body. There is a general duty on employers to restrict the exposure of employees and other people who could be affected by contamination. Restriction of exposure is achieved by engineering controls, design and safety features.

Confined Spaces

The **Confined Spaces Regulations 1997** require the full implementation of risk assessments, including safe systems of work, to avoid risks such as asphyxiation from gas, fume, or vapour when working in confined spaces.

Use of Risk Assessment in Development of Safe Systems of Work and Safe Operating Procedures

Systematic methodologies, such as task (or hazard) analysis, can be used to develop a safe system of work. One method is to use the following four-step approach:

1. Analyse the task.
2. Introduce controls.
3. Instruct and train.
4. Monitor and review the system.

Analysing the Task – Identifying the Hazards and Assessing the Risks

Assess all aspects of the task and the risks which it presents, considering hazards to health as well as to safety.

Take account of:

- What is used, e.g. the plant and substances, potential failures of machinery, toxic hazards, electrical hazards, design limits, risk of inadvertently operating automatic controls.
- Who does what, e.g. delegation, training, foreseeable human errors, short cuts, ability to cope in an emergency.
- Where the task is carried out, e.g. hazards in the workplace, problems caused by weather conditions or lighting, hazards from adjacent areas or contractors, etc.
- How the task is done, e.g. the procedures, potential failures in work methods, lack of foresight of infrequent events.

Where possible, you should eliminate the hazards and reduce the risks before you rely upon a safe system of work.

Introducing Controls and Formulating Procedures

Control selection may be from the range we identified earlier (technical, procedural, behavioural), taking account of the application hierarchy and the general principles of prevention. Your safe system of work may include verbal instructions, a simple written procedure or, in exceptional cases, a formal permit-to-work scheme. Other matters include:

- Considering the preparation and authorisation needed at the start of the job.
- Ensuring clear planning of job sequences.
- Specifying safe work methods.
- Including means of access and escape if relevant.
- Considering the tasks of dismantling, disposal, etc. at the end of the job.

Involve the people who will be doing the work. Their practical knowledge of problems can help avoid unusual risks and prevent false assumptions being made at this stage.

In those special cases where a permit-to-work system is necessary, there should be a properly documented procedure.

Instructing and Training People in the Operation of the System

Your safe system of work must be communicated properly, understood by employees and applied correctly. They should be aware of your commitment to reduce accidents by using safe systems of work.

Ensure that supervisors know they should implement and maintain those systems of work and that employees, supervisors and managers are all trained in the necessary skills and are fully aware of potential risks and the precautions they must adopt.

Stress the need to avoid shortcuts. It should be part of a system of work to stop work when faced with an unexpected problem until a safe solution can be found.

Monitoring and Reviewing the System

Safe systems of work are encouraged by training

Monitoring and reviewing involves periodically checking that:

- Employees continue to find the system workable.

- The procedures laid down in your system of work are being carried out and are effective.

- Any changes in circumstances which require alterations to the system of work are taken into account and implemented.

STUDY QUESTIONS

4. What is a permit to work?

5. Explain how risk assessment should be used to develop a safe system of work.

(Suggested Answers are at the end.)

Summary

Common Risk Management Strategies

We identified and discussed:

- Avoidance or elimination.
- Reduction.
- Transfer.
- Retention with/without knowledge.

Factors to Be Taken Into Account When Selecting Risk Controls

The **Management of Health and Safety at Work Regulations 1999** require control measures to be considered using the following principles:

- Avoid risks.
- Evaluate.
- Combat at source.
- Adapt work to the individual – ergonomics.
- Adapt to technical progress.

Control measures may be classified as being technical, procedural or behavioural.

The choice of control measures adopted should take account of:

- Use in the long or short term.
- Applicability.
- Practicability.
- Cost.
- Proportionality.
- Effectiveness.
- Legal requirements.
- Competence/training needs.

Safe Systems of Work and Permit-to-Work Systems

We considered the following in terms of **safe systems of work**:

- A safe system of work is one where the work is organised to remove hazards and minimise risks.
- It should proceed logically from identification to elimination of risks.
- A safe system of work is needed where hazards cannot be physically eliminated.

A **permit to work** is a formal written document of authority to undertake a specific procedure and is designed to protect personnel working in hazardous areas or activities.

Permit-to-work systems do not replace safe systems of work; they try to ensure that formal action is taken to eliminate **human error**.

When using risk assessment in the development of safe systems of work and safe operating procedures one method is to use the following four steps:

1. Analyse the task – identifying hazards and assessing risks.

2. Introduce controls and formulate procedures.

3. Instruct and train people in the operation of the system.

4. Monitor and review.

Exam Skills

The last question you attempted looked daunting but hopefully you will feel more confident at tackling this one. Here we will have a look at another of the short-answer, compulsory 10-mark questions.

Remember that with these Section A questions you have no choice. You need to average 50% across the six compulsory questions to give yourself a great chance of passing the exam. Most students find that there are questions they find easy to answer and others where their knowledge is weaker. You need to develop your skills and knowledge to raise your marks in your weaker questions, so make sure you have a go at all of them.

QUESTION

A health and safety management programme encompasses the following concepts:

(a) risk avoidance;	(2)
(b) risk reduction;	(2)
(c) risk transfer;	(3)
(d) risk retention.	(3)

Identify the key features of **EACH** of these concepts **AND give** an appropriate example in **EACH** case.

Approaching the Question

Note that your answer to this question will need to show a logical progression of ideas; think about the marks on offer for each part of the question.

HINTS AND TIPS

As you write your answer, make sure that you are identifying the right concept.

Suggested Answer Outline

(a) Risk avoidance – in this part of your answer, the examiner would have expected you to identify the first step to AVOID or ELIMINATE risk; you would need to give two or three examples of how you could do this. (Remember this is **avoiding**, not **substituting**.)

(b) Risk reduction – your answer should mention the evaluation of risks and the development of risk control strategies to achieve acceptable or tolerable levels of risk, using the hierarchy of control measures.

(c) Risk transfer – here you would explain the concept of transferring risk to other parties, but this could cost you in financial terms – examples might be insurance, use of contractors, business interruption recovery planning or outsourcing processes.

(d) Risk retention – you should mention accepting the residual level of risk and funding losses internally within the company. Also, point out that sometimes these risks will have been identified and accepted (known as 'risk retention with knowledge'); but sometimes they will be missed, forgotten or overlooked, leaving the company in a vulnerable position (known as 'risk retention without knowledge').

Example of How the Question Could Be Answered

(a) *Risk avoidance involves you either avoiding the risk or eliminating it completely. This isn't always easy to do, but you could mechanise a manual handling operation to eliminate it or avoid hot work by fabricating the equipment by fastenings or adhesives – the danger is you introduce another hazard you may not have considered.*

(b) *Risk reduction involves you deciding it isn't reasonably practicable to eliminate the risk, but you still have to control it and you need to apply the hierarchy of control measures, which would start with you reducing the hazard (extremely flammable for a flammable substance), isolating it (fixed guards of a machine with moving parts), controlling the risk (use of training, procedures, supervision, etc.) and as a last resort adopting PPE. The technique will involve you deciding what is an acceptable level of risk.*

(c) *Risk transfer involves passing risk onto someone else, i.e. insurance company, contractor, etc. but will involve you paying a premium for this service.*

(d) *Risk retention is the risks that the company takes on itself, either intentionally (they decide to live with the risk) or unintentionally where they have failed to identify the risk, which can put them in a vulnerable position.*

Reasons for Poor Marks Achieved by Candidates in Exam

An exam candidate would achieve **poor marks** for:

- Getting the different terms confused.
- Giving examples which did not match the concepts.
- Giving the same answer for different sections.

Organisational Factors

Learning Outcomes

Once you've read this element, you'll understand how to:

1 Explain the types of health and safety leadership, their advantages, disadvantages and likely impact on safety performance.

2 Explain the organisational benefits of effective health and safety leadership.

3 Explain the internal and external influences on health and safety in an organisation.

4 Outline the different types of organisation, their structure, function and the concept of the organisation as a system.

5 Explain the requirements for managing third parties in the workplace.

6 Explain the role, influences on and procedures for formal and informal consultation with employees in the workplace.

7 Explain health and safety culture and climate.

8 Outline the factors which can both positively and negatively affect health and safety culture and climate.

Contents

Contents

Types of Safety Leadership

IN THIS SECTION...

- Successful safety leadership requires active commitment by senior management, which is communicated downwards to all within the organisation.
- Transformational, transactional, servant, and situational and contextual are different styles of safety leadership.
- An effective leader is likely to have certain behavioural attributes.

Meaning of Safety Leadership

Hersey and **Blanchard** define leadership as:

"the process of influencing the activities of an individual or a group in efforts toward goal achievement in a given situation".

Source: *Management of organizational behavior*, Prentice Hall

Leadership is critical to achieving highest health and safety standard

Risk management is concerned with protecting the health and safety of employees or members of the public who may be affected by work activities. Since these activities are controlled and directed at board level, then their health and safety implications must be a board level issue as well. Failure to include health and safety as a key business risk in board decisions can have catastrophic results. This is illustrated by many high-profile safety incidents that have occurred over the years and in almost all cases the root cause is failure of leadership.

The legal framework places health and safety duties on organisations and employers. Members of the board therefore have both collective and individual responsibility for health and safety.

Successful safety leadership is based on visible, active commitment at board level, with effective downward communication systems through the management structure. The aim is to integrate good health and safety management with business decisions.

Effective leadership should involve the workforce in the promotion and achievement of safe and healthy conditions and encourage upward communication to engage the workforce. Without the active involvement of directors, organisations will never achieve the highest standards of health and safety management.

Types of Safety Leadership

There are a number of recognised theories relating to leadership style which can be associated with safety leadership.

Transformational

Transformational leadership is based on the assumption that people will follow a person who inspires them, and that the way to get things done is by generating enthusiasm and energy; consequently, the aim is to engage and convert the workforce to the vision of the leader. Since people will not immediately buy into radical ideas, the transformational leader must continually sell the vision and, as part of this, sell themselves. For this to work, transformational leaders need to have a clear idea of the way forward, and always need to be visible. This style is therefore a continuing effort to motivate the workforce.

Transformational leaders are people-oriented and believe that success is achieved through commitment, so the focus is on motivation and the involvement of individuals in the health and safety programme. However, the disadvantage of this approach is that passion and enthusiasm may not align with reality. The transformational leader may believe they are right, but this is only their belief. Transformational leaders are good at seeing the big picture – their vision – but sometimes not the detail where the problems often arise. They therefore need people to take care of things at this level.

Within the health and safety programme, transformational leaders focus on supervisor support, training and communication.

Transactional

Transactional leadership is based on the assumption that people are motivated by reward and punishment and social systems work best with a clear chain of command. The prime purpose of a subordinate is to do what their manager tells them to do, so the transactional leader creates clear structures setting out what is required and the associated rewards or punishments. The organisation and therefore the subordinate's manager has authority over the subordinate, and the transactional leader allocates work. The subordinate is fully responsible for it, whether or not they have the resources or ability to carry it out. When things go wrong, the subordinate is personally at fault, and is punished for failure. The assumption is that if something is operating to defined performance it does not need attention. Success requires praise and reward and substandard performance needs corrective action.

The style of transactional leadership is that of 'telling' in comparison to the 'selling' style of transformational leadership. It is a common approach for many managers but is closer to management rather than leadership.

The main **limitation** is the assumption that individuals are simply motivated by reward and exhibit predictable behaviour. However, this does not address the deeper needs identified in Maslow's hierarchy of needs (see next element).

Within the health and safety programme, transactional leaders focus on compliance, rules and inspection.

Servant

Servant leadership is based on the assumption that leaders have a responsibility toward society and those who are disadvantaged, so the servant leader aims to serve others and help them to achieve and improve. Key principles of servant leadership include personal growth, environments that empower and encourage service, trusting relationships to encourage collaboration, and the creation of environments where people can trust each other and work together.

Servant leadership puts the well-being of followers before other goals but could be seen as a weak leadership style. It may be viewed as an appropriate model for the public sector or large caring employers, but may be considered too caring and considerate for the private sector where the needs of shareholders, customers and market competition are more important. It also relies on the assumption that the followers want to change and serve others.

Within the health and safety programme, servant leadership focuses on co-operation, consultation, personal growth and well-being.

Situational and Contextual (Hersey and Blanchard)

Rather than promote a particular leadership style, **Hersey** and **Blanchard** recognise that tasks are different and each type of task requires a different leadership approach. A good leader will be able to adapt leadership to the goals to be accomplished. Consequently, goal-setting, capacity to assume responsibility, education and experience are identified as key factors that make a leader successful. As well as leadership style, the ability or maturity of those being led is also an important factor. Leadership techniques can therefore be optimised by matching the leadership style to the maturity level of the group, as follows:

- Leadership style:

  1. Telling – unidirectional flow of information from the leader to the group.

  2. Selling – the leader attempts to convince the group.

  3. Participating – the leader shares decision-making with the group, making the system more democratic.

  4. Delegating – the leader is still in charge, but monitors the ones delegated with the tasks.

- Maturity level of those being led:

  1. Incompetence or unwillingness to do the task.

  2. Inability to do the task but willing to do so.

  3. Competent to do the task but not confident.

  4. Competent and confident.

So, for example, where followers lack competence they need direction and supervision (**telling**) to get them started; but where followers are competent and confident they are able and willing to work by themselves with little supervision or intervention. Consequently, the leader only needs to provide such followers with clear objectives and some limits to their authority, but otherwise let them get on with it (**delegating**).

Comparison of Leadership Styles

The advantages and disadvantages of each of the different leadership styles can be summarised as follows:

Leadership Style	Advantages	Disadvantages
Transformational	Promotes two-way communication Creates strong bond between manager and employee Encourages continual improvement Helps employees adapt to changes Encourages work ownership Encourages employees to become more active so develops next generation of leaders	Leader's passion and enthusiasm may not align with reality Encourages concentration on the big picture so may lose sight of the detail
Transactional	Encourages consistent quality processes and outcomes Not dependent on personal traits such as charisma or inspiration Leaves little room for misinterpretation or ambiguity Works well when short-term results are needed fast	Too simplistic – fails to account for individual motivations Unwillingness to consider other ideas limits leader's ability to adjust if things go wrong Employees may become unhappy and dissatisfied Leader must be present to guarantee the work gets done properly (Continued)

Leadership Style	Advantages	Disadvantages
Servant	Allows for personalised management for each member of team Develops sense of loyalty from employee to company Gets employees involved in decision-making Encourages high sense of morale which can help increase productivity	May be seen as a weakness May be 'too soft' for the private sector Relies on the assumption that followers want to serve others
Situational/ Contextual	Easy to understand and use Employees are given appropriate level of direction and support based on individual needs Flexible	Managers must accurately be able to assess the employee's maturity and skill level Can result in inconsistency (employees may not know what to expect) Can be perceived as manipulative or coercive

Behavioural Attributes of an Effective Leader

The styles of leadership we have discussed suggest a broad spectrum between the two extreme approaches of **autocratic** and **democratic**. However, there are also leadership **behaviours** which are regarded as being effective and are respected by followers.

These include:

- Integrity.
- Appreciation of corporate responsibility (the need to make profit is balanced with wider social and environmental responsibilities).
- Being emotionally positive and detached.
- Leading by example.
- Supporting and backing people when they need it.
- Treating everyone equally and on merit.
- Being firm and clear in dealing with bad behaviour.
- Listening to and understanding people ('understanding' is different to 'agreeing').
- Always taking responsibility and blame for mistakes and giving people credit for successes.
- Being decisive and seen to make fair and balanced decisions.
- Asking for views, but remaining neutral and objective.
- Being honest but sensitive in delivering bad news or criticism.
- Keeping promises.
- Always accentuating the positive.
- Involving people in thinking and especially in managing change.

STUDY QUESTION

1. Outline the basic principles of the following types of health and safety leadership:

   (a) Transformational.

   (b) Transactional.

   (c) Servant.

(Suggested Answer is at the end.)

Benefits of Effective Health and Safety Leadership

IN THIS SECTION...

- The Health and Safety Executive (HSE)/Institute of Directors (IOD) has issued guidance for the effective leadership of health and safety which includes:
  - Active leadership.
  - Worker involvement.
  - Assessment and review.
- Effective health and safety management is significantly influenced by appropriate leadership; strong management commitment will benefit the organisation's health and safety culture and performance.
- Leadership plays an essential role in promoting participation and engagement of the workforce in effective health and safety management.
- Both the health and safety practitioner and the organisation have a leadership role in the achievement of high standards of health and safety in the workplace.
- Social corporate responsibility refers to the voluntary actions that businesses undertake to address not only their own needs but also those of the wider society as they relate to health and safety management.
- The Financial Reporting Council has established guidance for assessing the effectiveness of risk control measures.

Purpose of the HSE/IOD Guidelines 'Leading Health and Safety at Work'

> **MORE...**
>
> You can find further details of the IOD's guidelines at:
>
> www.iod.com/influencing/policy-papers/regulation-and-employment/leading-health-safety-at-work
>
> Or download the leaflet from the HSE website at:
>
> www.hse.gov.uk/pubns/indg417.pdf

The purpose of this guidance is to set out an agenda for the effective leadership of health and safety and is designed for use by all directors, governors, trustees, etc. in organisations of all sizes.

The following essential elements are identified:

- Strong and active leadership from the top:
  - Visible, active commitment from the board.
  - Establishing effective 'downward' communication systems and management structures.
  - Integration of good health and safety management with business decisions.

- Worker involvement:
  - Engaging the workforce in the promotion and achievement of safe and healthy conditions.
  - Effective 'upward' communication.
  - Providing high-quality training.
  - Assessment and review:
  - Identifying and managing health and safety risks.
  - Accessing (and following) competent advice.
  - Monitoring, reporting and reviewing performance.

Leadership as a Core Element of Effective Health and Safety Management

Organisations have management arrangements to deal with personnel, finance and quality control so health and safety should be considered no differently. Managing health and safety needs to be an integral part of the everyday process of running an organisation.

The core elements to effectively managing for health and safety rely on:

- Leadership and management.
- A trained and skilled workforce.
- An environment where people are trusted and involved.
- An understanding of the risk profile of the organisation.

To achieve effective health and safety management in an organisation, leaders at all levels need to understand the range of health and safety risks in their part of the organisation and to recognise their importance. This means involvement in assessing risks, implementing controls, supervising and monitoring.

Leadership is a core element of effective health and safety management

When board members do not lead effectively on health and safety management, the consequences can be severe.

In practice, effective leadership involves:

- Maintaining attention on the significant risks and implementation of adequate controls.
- Demonstrating commitment by actions and awareness of the key health and safety issues.
- Consulting with the workforce on health and safety.
- Challenging unsafe behaviour in a timely way.
- Setting health and safety priorities.
- Understanding the need to maintain oversight of the risks and controls.
- Showing acceptance and compliance with the organisation's standards and procedures (e.g. wearing the correct Personal Protective Equipment (PPE) on site).
- Striving to engage employees in the health and safety programme.

Benefits of Effective Safety Leadership on the Health and Safety Culture and Performance of an Organisation

Achieving a positive health and safety culture in an organisation is fundamental to managing health and safety effectively and leaders can influence this by:

- Understanding the effect of different levels of management on the organisational culture.

- Making sure that all managers are committed to promoting health and safety.

- Recognising that the attitudes and decisions of senior managers are critical to the culture of the organisation and in setting priorities.

- Encouraging a leadership role in managers so that they are not simply restricted to directing work and monitoring compliance with rules and regulations, but act as facilitators and engage with the workforce to solve health and safety problems.

- Recognising the important part that employees play in shaping the safety culture of the organisation and engaging with them to encourage joint involvement in the health and safety programme.

- Making sure that health and safety is not viewed as a separate function but as an integral part of the business, and that health and safety risks are recognised as part of key business risks.

The tangible benefits of a positive health and safety culture are reflected in indicators of good health and safety performance and include:

- Reduced costs.

- Reduced risks.

- Lower employee absence and turnover rates.

- Fewer accidents.

- Lessened threat of legal action.

- Improved standing among suppliers and partners.

- Better reputation for corporate responsibility among investors, customers and communities.

- Increased productivity, because employees are healthier, happier and better motivated.

Link Between Effective Leadership and Employee Engagement

Employee consultation and involvement is an essential element of effective health and safety management so leadership plays an essential role in promoting participation and engagement of the workforce.

The legal requirements for consultation and involvement of the workforce include:

- Providing information, instruction and training.

- Engaging in consultation with employees, and especially trade unions where they are recognised.

Beyond the required legal minimum standard, worker involvement can extend to full participation of the workforce in the management of health and safety. This serves to create a culture where relationships between employers and employees are based on collaboration, trust and joint problem solving. Employees are involved in assessing workplace risks and the development and review of workplace health and safety policies in partnership with the employer.

Effective health and safety leadership will ensure that:

- Instruction, information and training are provided to enable employees to work in a safe and healthy manner.

- Safety representatives and representatives of employee safety carry out their full range of functions.

- The workforce is consulted (either directly or through their representatives) in good time on issues relating to their health and safety and the results of risk assessments.

- Employees are clear who to go to if they have health and safety concerns.

- Line managers regularly discuss how to use new equipment or how to do a job safely.

- Health and safety information is cascaded through the organisation through team meetings, noticeboards and other communication channels.

In order to achieve employee engagement, effective leaders need to recognise that:

- The health, safety and well-being of the workforce is paramount and employee participation has an essential contribution to make.

- Successful businesses increasingly encourage active participation of the workforce in the management of health and safety.

- Involving staff in the process of identifying and managing risks is a key aspect of managing health and safety successfully.

- Reviewing progress against agreed objectives at regular intervals, setting performance measures and developing an improvement plan provides evidence to the workforce of continual improvement.

Encouraging Positive Leadership for a Safe and Healthy Workplace

We have already demonstrated the need for effective health and safety leadership in order to achieve high standards of health and safety in the workplace. Both the safety practitioner and the organisation as a whole have a role to play in this.

Health and Safety Practitioner

One function of the health and safety practitioner is to advise on aspects of the health and safety management programme, such as:

- Formulating and developing health and safety policies and plans.

- Profiling and assessing risks and organising activities to implement the plans.

- Measuring performance.

- Reviewing performance and taking action on lessons learnt.

For the programme to be effective, it needs leadership and therefore the health and safety practitioner is required to motivate the board into action in order to:

- Set the direction for effective health and safety management.

- Establish a health and safety policy that is an integral part of the organisation's culture and values.

- Take the lead in ensuring the communication of health and safety duties and benefits throughout the organisation.

- Respond quickly where difficulties arise or new risks are introduced.

Health and safety practitioners can not only contribute to the achievement of the objectives of an organisation by encouraging leadership, but also by leading on health and safety issues themselves. They can act as advocates, persuading both management and the workforce of the value of their knowledge and expertise. In organising activities to deliver the programme, the health and safety practitioner has a role in involving all levels of management and the workforce and communicating so that everyone is clear on what is needed. The health and safety practitioner can discuss issues raised and lead in developing positive attitudes and behaviours.

The Organisation

Policy development and planning is a key component of the organisation's health and safety management system. This starts with a statement of intention which establishes how a safe and healthy environment for the workforce (and anyone else who could be affected by work activities) will be achieved. The organisational structure to attain this sets out everyone's roles and responsibilities including directors, managers, supervisors and workers. The arrangements clarify how things will be done, including details of the systems and procedures needed to meet legal obligations. In addition, measurement of health and safety performance needs to be established along with agreed performance targets.

This policy and planning process should serve to commit leaders to an effective health and safety programme. Leaders should respond by demonstrating commitment and leading by example through:

Policy and planning commits leaders to an effective health and safety programme

- Showing that health and safety is an important issue reinforced by visible action.

- Promoting health and safety whenever possible to a wide range of audiences.

- Discussing health and safety early in the agenda of every management meeting.

- Ensuring that health and safety is a significant element of performance reviews.

- Setting longer-term health and safety goals to show commitment to continual improvement.

- Holding line management and staff accountable for health and safety but not looking to apportion blame.

- Ensuring that the organisation routinely reports on health and safety performance as part of a commitment to Corporate Social Responsibility (CSR).

- Making sure that managers know there are adequate resources to work in a healthy and safe manner and that 'corner cutting' on health and safety standards will not to be tolerated.

- Measuring health and safety performance through useful and meaningful indicators which compare performance both internally over time, and also externally against others working with similar hazards.

- Setting long-term goals for the control of major hazards as is done for financial or production goals.

- Meeting with the workforce regularly to discuss health and safety and encouraging staff to raise health and safety concerns.

- Ensuring that all incidents and near misses are investigated fully to identify the underlying causes and establish effective remedial action.

Corporate Social Responsibility

This is the term used to describe the voluntary actions that business can take, over and above compliance with minimum legal requirements, to address both its own competitive interests and the interests of the wider society. Businesses should take account of their economic, social and environmental impacts, and act to address the key sustainable development challenges based on their core competences wherever they operate – locally, regionally and internationally.

Companies are being put under increasing pressure to measure and report on health and safety issues through their CSR policies. This is because occupational health and safety, as well as product safety, is now widely recognised to form an integral part of CSR and is included in all major measurement and reporting guidelines and tools developed for CSR. Organisations are no longer simply reporting financial performance data.

The pressure from shareholders, investors and other stakeholders to improve CSR and run a business ethically and transparently, not only enhances reputation but leads to improvements in health and safety.

Influence of the Financial Reporting Council Guidance on Internal Control

The Institute of Chartered Accountants published *Internal control: Guidance for directors on the Combined Code* in 1999. It is commonly called 'The Turnbull Report' because the relevant committee was chaired by Nigel Turnbull. This report consisted of a set of recommendations outlining the fundamental need for risk management, i.e. the use of a risk-based approach for internal control. These recommendations became mandatory in December 2003.

In 2004, the Financial Reporting Council (FRC) established the Turnbull Review Group which provided some updating to the guidance.

The appendix to the report includes a useful checklist for assessing the effectiveness of a company's risk and control processes. The Institute of Chartered Accountants has also produced a boardroom briefing, *Implementing Turnbull*, which contains practical advice and case studies.

The FRC Report requires:

- **Clear Policies and Commitment**

    The Board of Directors should set a clear policy on risk and internal control. All levels of the company need an understanding of it and should be committed to implementing it.

- **Risk Assessment**

    This involves identifying the significant business risks and evaluating their significance (prioritising). Like health and safety risks, business risks can be described by reference to their impact on the business (severity) and the likelihood of their occurrence. Risk can be evaluated before consideration of existing control measures and also after their inclusion (residual risk).

- **Control Environment and Control Activities**

    Each business must have a clear strategy for dealing with significant risks. The company's culture should support the business objectives and risk management. Authority, responsibility and accountability should be clearly defined and there should be effective communication.

- **Clear Communication and Reporting Arrangements**

    Clear channels of communication should be established. This is necessary for the periodic reporting of progress (with respect to business objectives and related risks). This should also include mechanisms for reporting suspected breaches of laws/policies.

- **Monitoring and Auditing**

    Effective monitoring processes are required. These might include such things as statements of compliance with policies and a code of conduct from employees. Internal audits of compliance could be done by a separate function within the company, independent of line management.

    The Board should perform an annual assessment/review of all aspects of its internal control processes. This is a prerequisite for making its annual public statement to shareholders on internal control.

In summary, the FRC Report requires companies to identify, evaluate and manage their significant risks and to assess the effectiveness of the related internal control systems. The board of directors must review the effectiveness of their internal control system and undertake an annual assessment of it in order to make a statement regarding internal control within the company's annual report.

In essence, the point and purpose of these recommendations is to ensure that directors take responsibility for adequate risk management. If they do not, they must disclose the fact to their shareholders, and risk the effect this may have on the reputation of their company and the stock market.

While the FRC Report originates from a financial institute, the concepts raised should be familiar to safety professionals as they appear in a number of safety management systems. As such, the 'Turnbull Report' recommendations are supportive of safety management systems such as HSG65, which we discussed earlier.

STUDY QUESTIONS

2. Outline tangible benefits of a positive health and safety culture.

3. Outline how effective leadership can play an essential role in promoting participation and engagement of the workforce.

4. The 'Turnbull Report' proposed recommendations for the use of a risk-based approach for internal control. Outline the main elements of this approach.

(Suggested Answers are at the end.)

Internal and External Influences

IN THIS SECTION...

- The key internal influences on health and safety are finance, production targets, trade unions, and organisational goals and culture.

- The key external influences include legal issues (legislation, Parliament and the HSE, enforcement agencies, courts and tribunals, and contracts) and third parties (clients and contractors, trade unions, insurance companies and public opinion).

Internal Influences on Health and Safety Within an Organisation

Finance

Setting up and running a company requires considerable financial investment. Once established, the company needs to generate more income than it spends on running costs, i.e. cost of premises, plant, wages, insurance, etc. To do this, the company will set annual budgets specifying the amount of money available to each department to support its running costs and setting production targets to be achieved. When budgets are being reduced to economise, some health and safety requirements will often be 'short circuited'. The person responsible for health and safety must argue for sufficient funds to support health and safety requirements. Lack of funding will inevitably lead to a reduction in the resources necessary to effectively administer health and safety. Health and safety costs might seem to be minimal and easily absorbed in departmental administration costs. Such an arrangement could lead to financial disaster and costly prosecutions for non-compliance.

Internal influences include finance and production targets

Production Targets

Achieving production goals can put intense pressures on workers leading to stress and an increase in incidents and accidents in the workplace. It is recognised that increased competition, longer hours, increased workloads, new technology and new work patterns are significant occupational stressors. Industrial psychology also requires that in a 'conveyor-type' operation, the speed of the belt should be geared to the capacity of the slowest operator. The pressures on management to achieve production targets/increase production can be translated into action on the shop floor in a number of ways:

- Make the workforce work longer hours.

- Increase the size of the existing workforce.

- Pay incentive bonuses to increase the daily rate of production.

- Reduce the quality of the goods by using inferior materials.

Apart from increasing the size of the workforce, these measures encourage workers to 'cut corners'. For example:

- Longer hours can lead to tiredness and less attention to safety factors.

- Bonuses for increased production can lead to disregard for any safe systems of work which slows down the speed at which the worker can operate.

- Increased production targets may create anxiety in the slower worker, especially if part of a team, and can lead to shortcuts being taken in an effort to keep up with colleagues.

- Reducing quality may require new systems of work, leading to stress.

All of these can lead to unsafe acts that may have a considerable effect on the company's health, safety and accident record.

Trade Unions

Trade union safety representatives are involved as members of safety committees and, as such, are actively involved in improving health and safety in the workplace. They have a dual role in that they can be involved in the formulation of policy in certain companies, but they also have a policing role in that they can monitor management's performance. They carry out the following functions:

- Investigating potential hazards and dangerous occurrences.

- Examining the cause of accidents.

- Investigating health and safety complaints from employees they represent.

- Making representations to the employer on complaints, hazards and accidents.

- Carrying out inspections of the workplace.

- Consulting with HSE inspectors on behalf of the employees they represent.

- Receiving certain information from the HSE inspector.

Employee representation has been widened to include employees who are not members of a trade union. These employees will be represented by 'elected representatives of safety'. Safety representatives are protected by legislation from victimisation by employers.

Organisational Goals and Culture

The goals and culture of the organisation strongly characterise the company. Some organisations rate safety highly and treat it seriously, not only in what they claim to do (their safety policy), but also in what actually happens in practice. Safety culture can be simply described as 'the way we do things'. If you have worked for several different organisations you will probably recognise different cultures in terms of what they accept and tolerate. We will look at this topic in more detail later in this element.

External Influences on Health and Safety Within an Organisation

The following all exert an influence on the organisation (positive or negative):

External influences on the organisation

Legislation

Any company ignores legislation at its peril. Changes in legislation are well-publicised in the appropriate publications and any health and safety adviser should ensure that they are aware of any pending changes and their effect on the company.

Parliament/HSE

Of all the influences on a company, probably the most important is that of legislation. The laws passed by governments will have a direct effect on any company and changes in procedures to accommodate legislative changes may be necessary.

The HSE can create change by publishing Approved Codes of Practice which recommend good practice. While these do not have the force of law, companies must show that they have adopted a standard at least equal to that published in the Code.

Enforcement Agencies

The enforcement agencies can influence health and safety within companies by:

- Providing advice.

- Serving Improvement and Prohibition Notices.

- Prosecution.

Tribunals/Courts

Employment tribunals may have a direct effect on health and safety through their decisions, such as dismissing an appeal against an Improvement Notice.

In a criminal prosecution, the court establishes whether the defendant is guilty or not guilty. The defendant may be an individual or the company itself. If the prosecution is successful, the organisation will in most cases be fined.

In civil cases for personal injury, the organisation may be sued, which may result in compensation being paid to the injured party.

Contracts/Contractors/Clients

The nature of contracts and relationships with contractors may have profound effects on the health and safety of a particular contract. Where a contractor feels that they are making a loss on a particular project, there may be a strong temptation to cut corners and perhaps compromise on health and safety. Where a client takes a direct interest in the progress of a contract and in achieving good standards of health and safety, the standards on site are positively improved. There is a need for effective vetting of contractors' own company health and safety competence before hiring their services.

Trade Unions

Trade unions are active nationally in promoting standards of health and safety in many ways:

- Supporting their members' legal actions and setting precedents and standards.

- Acting through lobby and pressure groups to influence legislation.

- Carrying out and sponsoring research.

- Publicising health and safety matters and court decisions.

- Providing courses on health and safety subjects.

Insurance Companies

Insurance companies directly influence other companies by means of the requirement for employers' liability insurance. Should a company suffer an unusually high accident rate, then the insurance company can either increase their insurance premiums or insist that the company adopt risk-reduction measures. Insurance companies now often carry out their own inspections of workplace risks and so are able to set certain minimum standards.

Insurance companies may also affect companies by means of their policy toward claims, i.e. because of the high cost of litigation cases tend to be settled out of court, rather than pursued in court.

Public Opinion

Ultimately, public opinion can have a powerful effect on legislators, which may result in legislation being passed or prosecution taking place. Pressure groups may lobby Parliament and influence the government to change the law. Following a series of major rail crashes in the late 1990s, survivors and relatives formed a group to try to force the government to improve safety standards on the railways and to hold the railway companies more accountable.

STUDY QUESTIONS

5. List some of the internal influences on an organisation in respect of health and safety at work.

6. List some external bodies that can influence health and safety standards of organisations, identifying the means by which each body exerts its influence.

(Suggested Answers are at the end.)

Types of Organisations

IN THIS SECTION...

- An organisation may be considered to be a system that has interacting components forming a whole.
- Within an organisation, there are both formal and informal structures.
- Conflict may arise as a result of individual goals not being consistent with those of the organisation.

Concept of the Organisation as a System

DEFINITION

SYSTEM

A regularly interacting or interdependent group of items forming a united whole.

(Note: This is one of several definitions which can be applied to systems.)

The systems approach to management is a way of thinking in which the organisation is viewed as an integrated complex of interdependent parts which are capable of sensitive and accurate interaction among themselves and within their environment.

Common **characteristics of systems** are that:

- Every system is part of a still larger system and, itself, encompasses many subsystems ('circles within circles').
- Every system has a specific purpose to which all its parts are designed to contribute.
- A system is complex – any change in one variable will effect change in others.
- Equilibrium: a system strives to maintain balance between the various pressures affecting it, internally and externally. Some systems experience more pressures to change than others, giving rise to stable and unstable systems.

Initial reaction to pressure is often what is called **dynamic conservatism** – the organisation fights like mad to stay just as it is! However, sooner or later, homeostasis takes place (activities that serve to stabilise and vitalise the organisation as a whole in an evolving state of dynamic equilibrium).

Organisational Structures and Functions

General Perspective

An organisation is a group of persons who interact with each other in an effort to achieve certain goals or objectives. At a very basic level, the shop-floor employee goes to work to earn money – as does their union representative, foreperson, manager and managing director. The earning of money, then, is a specific goal common to everyone in that particular enterprise. There will be many other shared goals and objectives as well as many goals that are not shared, which lead to conflict, and that may eventually have a bearing on the success or failure of the organisation. A work organisation, then, is an organisation which has been established for a specific purpose and within which work is carried out on a regular basis by paid employees. Examples of such are: businesses, hospitals, educational institutions, government departments, etc.

Formal and Informal Structures

All organisations have a formal and informal structure. Within each organisation, there is a formal allocation of work roles and the administrative procedures necessary to control and integrate work activities.

However, organisations also have an informal arrangement or power structure based on the behaviour of workers – how they behave toward each other and how they react to management instructions. The foreperson or supervisor will have specific instructions from management aimed at achieving certain goals or production targets. In many cases, they 'adjust' those instructions in accordance with their personal relationships with individual, or groups of, workers. This takes us some way toward being able to make a distinction between formal and informal organisations. There is a blurring at the edges between the two – a cross-over point where the distinction between the formal and informal at the actual point of action becomes obscured and is the subject of a great deal of sociological argument and discussion. For our purposes, we can describe or explain them in the following way:

- **Formal Organisational Structure**

    Most organisations describe their structure in the form of an organogram. This shows the reporting relationships, from the chief executive of the company down to the staff carrying out the most basic tasks.

    The following figure illustrates a typical formal structure for a small company.

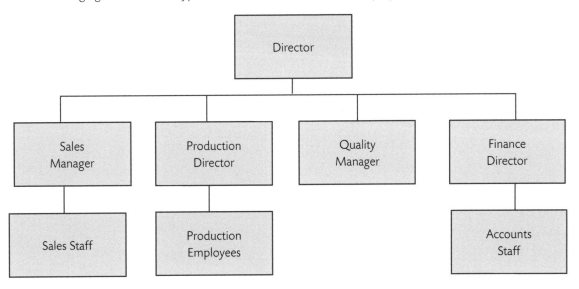

Formal structure

    In theory, every person within the structure has a well-defined role with clear lines of reporting and clear instructions about the standards of performance. These roles are clearly understood by others in the organisation so that everyone acts together to achieve the organisational objectives.

- **Informal Organisational Structure**

    An organisational chart cannot identify all the interactions that occur between staff. Invariably, it will be the quality of personal relationships which determines how communications flow within a company and 'how things get done'.

    In most organisations, the formal structure represents the model for interaction, but, in reality, the informal relationship is significant in understanding how organisations work. The informal structure cannot **replace** the formal structure, but works **within it**. It can influence relationships and effectiveness in both positive and negative ways. An understanding of it is an invaluable aid to good management. Take another look at the **formal structure** figure and then compare it with the **informal structure** figure that follows. Look at the superimposed informal structure shown by the dotted lines.

An awareness of these informal relationships would obviously influence how communications are made. The effective manager will use such knowledge to break down resistance to new measures (including health and safety).

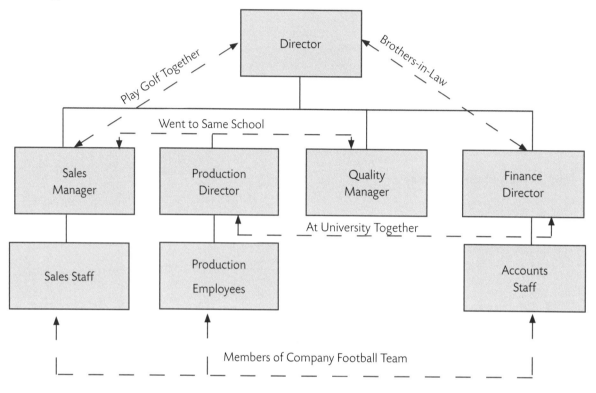

Informal structure

A simple way of making a distinction between formal and informal organisation structure is:

- **Formal** – represented by the company organisation chart, the distribution of legitimate authority, written management rules and procedures, job descriptions, etc.

- **Informal** – represented by individual and group behaviour.

Organisation Charts

The structure of an organisation is determined by its general activities – its size, location, business interests, customer base, etc. and by the way in which its employees are organised.

The organisational pyramid (formal structure) illustrated earlier is probably the principle model for most organisations, with management at its apex and the workforce at its base. Within this model, each separate department has its own pyramid with its own power structure and departmental goals. If the organisation is very large then considerable problems involving communication, efficiency, effectiveness, etc. may occur. The following figures show two typical pyramids.

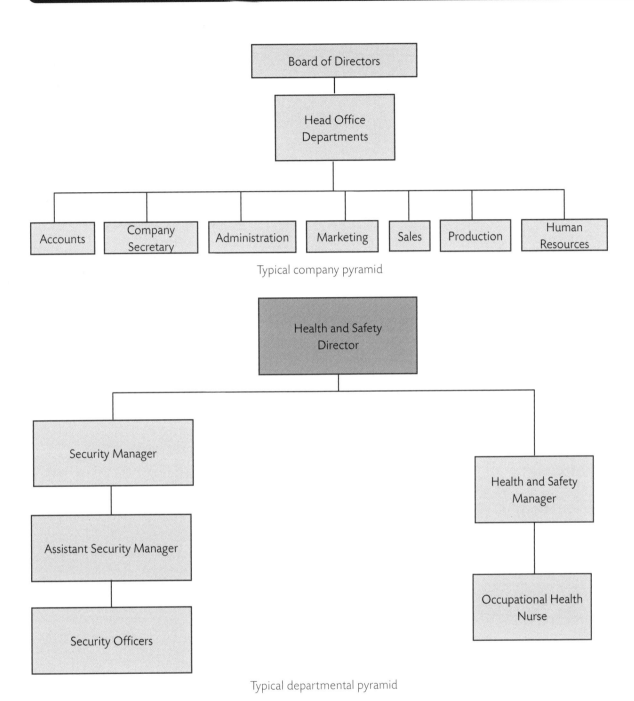

Typical company pyramid

Typical departmental pyramid

By looking at these structures, you can see the formal levels of authority and responsibility within the organisation or department. In simple terms, authority or control runs from top to bottom. However, there are other important management/employee relationships, such as line management, staff, and functional relationships.

Role of Management

Management will lead through issued instructions, policies and procedures, and supervision to ensure that these are being adhered to.

There is normally a line of responsibility with different functions at each level.

Hierarchical Line Management Structures

DEFINITIONS

MANAGER

A manager organises, regulates and is in charge of a business.

MANAGEMENT

The management are those people engaged in these functions.

Look at the following figure:

Works Director

Works Manager

Foreperson

Chargehand

Shop-Floor Operative

A typical line management function

Here you can see a direct line of authority from the works director to the shop-floor operative.

Staff Relationship

The Managing Director's (MD's) secretary reports to the MD and carries out instructions by passing the MD's wishes to other directors and senior heads of department, but there is no 'line' relationship between the secretary and those departments. There is no instruction from the secretary, as their authority stems from the MD. A health and safety consultant reporting directly to an MD is not in a position to 'instruct' heads of departments to carry out health and safety policies or instructions. Again, their authority stems from the MD and, in practice, they would advise heads of department of any changes in policy agreed with and authorised by the MD.

Functional Relationship

In many larger organisations, certain members of staff have a company-wide remit to carry out activities 'across the board'. Human resources departments often implement company appraisal plans which affect every department, internal auditors visit all departments to carry out their work, and quality control inspectors and health and safety managers have a company-wide role in order to inspect and check procedures. In such circumstances, any defects discovered would normally be dealt with by reporting them to the departmental head rather than dealing directly with any individual within the department.

The various hierarchies and line, staff and functional relationships can create huge problems for any organisation. Office 'politics' and protocols often obstruct communication, which is one of the keys to efficient management.

Small Businesses and Flat Management Structures

These are organisations with up to 50 employees. A feature of such organisations is the necessity (certainly in those with few employees) for the employees to adopt several roles. Much of the work is done in teams where a team leader will facilitate the work of the team, operating in a collaborative style rather than through a hierarchical structure. This is a much flatter structure than the linear one and relies on co-operation and joint decisions rather than instruction being passed down through a management chain.

Small businesses are far less likely to have a dedicated health and safety professional than a large organisation; the role is often taken on by an employee who combines the responsibility with other tasks.

Organisational Goals and Those of the Individual: Potential Conflict

To be successful and progress, both an organisation and individuals have to have goals. For the organisation, the goal may be an objective to be the 'best in their field' or to be the 'largest' or to be renowned for 'outstanding quality'. For the organisation to achieve these goals, the employees need to have their own goals and objectives to work toward the organisational goal. However, the individual may have other goals which may or may not impact on the organisation. For example, an individual may hope to be promoted, which would probably mean that they will work very hard to achieve their goals/objectives within the organisation as this should help them to achieve their own personal goal of promotion. Another individual, however, may want to work fewer hours or have more time with their family, and this may impact negatively on their willingness to put in extra hours which may be required for the organisation to achieve its goal.

DEFINITION

GOAL

An object of effort or ambition.

Integration of Goals of the Organisation with the Needs of the Individual

In setting and achieving health and safety targets, the organisation should consider the needs of the individual. Where health and safety tasks are delegated, at all levels from senior managers to shop-floor workers, the **responsible** individual(s) should be clearly identified and stated. This gives ownership to the individual concerned, and is an important factor in getting the individual to 'buy in' to the organisation's goals.

Many organisations give responsibility without the relevant **authority** to carry out the tasks. This can be a mistake as, without authority, the individual can feel frustrated at being unable to carry out the tasks. This leads to a feeling of futility and results in tasks being done poorly or not at all. Where authority is given to enable the individual to carry out tasks, this can result in an increase in self-esteem and every chance that the tasks will be performed well.

Individuals should be given ownership of health and safety tasks

The limits of responsibility and authority should be clearly defined so that individuals know the extent of what they can and cannot do.

With responsibility comes **accountability**, and this must be made clear to all individuals given health and safety responsibilities. One important issue when giving responsibility is to ensure that the individual is capable of accepting it.

STUDY QUESTION

7. What is the difference between a formal and informal organisational structure?

(Suggested Answer is at the end.)

Requirements for Managing Third Parties

IN THIS SECTION...

- The main third parties (i.e. non-employees) that need to be considered are contractors, agency workers and other employers (shared premises).

- When using contractors, certain procedures need to be adopted:

  - Planning – including risk assessment.

  - Selection – competent contractor.

  - During contract – ensuring contractor is inducted and is aware of local procedures.

  - Checking performance.

- Review procedures.

Third Parties

A third party is defined as 'someone other than the principals who are involved in a transaction'. The significance of this in the workplace is to distinguish between employees, for whom the employer has well-defined legal responsibilities, and others who may be affected by the safety of a workplace.

In a workplace it should be relatively straightforward for employers to discharge their responsibility for the employees on their payroll. However, it may be the case that the employer is operating in shared premises and therefore needs to consider the effects of their activities on **others sharing the premises**. In addition, as well as permanent employees under a contract of employment on the payroll, business needs may require the use of **temporary workers** on short-term contracts or contractors on a specific contract for a particular job.

Other Employers (Shared Premises)

Where employers share workplaces, they need to co-operate with each other to comply with their respective health and safety obligations. This involves telling other employers about any risks their work activities could present to their employees, both on- and off-site. The **Management of Health and Safety at Work Regulations 1999 (MHSWR, Regulation 11)** places a specific duty on all employers sharing a workplace or site to 'co-operate and communicate' with each other with the principal objective of ensuring the health and safety of all working at and visiting the premises. However, it can be difficult to establish exactly who is responsible for what, which is why communication and co-ordination is required and the respective employers must decide this for themselves.

The main principle that applies is that employers will be responsible for those activities and issues that are under their control, but co-operation and communication with others will still be required. As might be expected, the starting point for all parties is risk assessment, which needs to consider the risks to others sharing the building or site.

DEFINITIONS

CONTRACTOR

Someone who is engaged to perform a certain task without direction from the person employing them.

AGENCY WORKER

An individual who has a contract with a temporary work agency and who is supplied by that agency to work temporarily under the supervision and direction of the hirer.

Agency Workers

There is increasing use of agency workers, employed on a temporary basis, to supplement the labour force. Businesses and self-employed people using temporary workers must provide the same level of health and safety protection for them as they do for employees. Providers of temporary workers and employers using them need to co-operate and communicate clearly with each other to ensure risks to those workers are managed effectively. Again, it needs to be agreed who does what in this respect. If it is assumed that the 'other party' will take responsibility then workers may be left without any health and safety consideration or protection at all.

Before temporary workers start, they need to be covered by risk assessments and to know what measures have been taken to protect them. They also need to understand the information and instructions required for them to work safely and be provided with the necessary training. There may also be issues regarding language needs of temporary workers who do not speak English well or at all.

Other relevant issues include:

• The need to check on occupational qualifications or skills needed for the job.

• Agreement on arrangements for providing and maintaining any PPE.

• Agreement on arrangements for reporting accidents to the enforcing authority.

Contractors

Employers who engage contractors have health and safety responsibilities for the contractors and also for anyone else that could be affected by their activities. In addition, contractors themselves have legal health and safety responsibilities as employers or as employees. Again, co-operation and co-ordination are important to make certain that everyone understands the part they need to play to ensure health and safety.

Poor management of contractors can lead to injuries, ill health, additional costs and delays so it is important to work closely with the contractor in order to reduce the risks to employees and the contractors themselves.

Contractors can be at particular risk because they may be strangers to the workplace and unfamiliar with local procedures, rules, hazards and risks. The level of control needed over contractors depends on the complexity and the degree of risk associated with the task.

Monitoring the work of a contractor

Internal Rules and Procedures Concerned with the Selection, Appointment and Control of Contractors

To ensure that a chosen contractor is capable of doing the work required safely, you need to introduce procedures that will identify and cover key points. The following lists suggest an approach that covers all aspects of contractor hire (after HSE guidance HSG159 *Managing contractors*).

Planning

• Define the task(s) that the contractor is required to carry out.

• Identify foreseeable hazards and assess the risks from those hazards.

• Introduce suitable control measures to eliminate or reduce those risks, including permit-to-work systems if necessary.

- Lay down health and safety conditions specific to the tasks.
- Involve the potential contractors in discussions concerning the health and safety requirements.
- Agree realistic timescales for the work.
- Ensure accident and first-aid procedures and arrangements are in place.

Choosing a Contractor

- Determine what technical and safety competence is required by the contractor.
- Ask the contractor to supply evidence of that competence.
- Supply information regarding the job and the site, including site rules and emergency procedures.
- Ask the contractor to provide a safety method statement outlining how they will carry out the job safely.
- Decide whether sub-contracting is acceptable and, if so, how the main contractor will manage the health and safety of sub-contractors.

TOPIC FOCUS

Factors to be considered when selecting a contractor include:

- Experience/reputation/references.
- Quality of their health and safety policy.
- Quality of sample risk assessments.
- Training and qualifications of employees and safety advisers.
- Accident and enforcement history.
- Maintenance of equipment, including statutory inspections.
- Monitoring and consultation arrangements.
- Sub-contractor selection and control procedures.
- Suitability of detailed risk assessments and method statements.

Contractors Working on Site

- Introduce a signing-in-and-out procedure.
- Provide a named site contact who will go over the job with the contractors.
- Ensure contractors are inducted.
- Pass on information about the site regarding the hazards and risks, site rules and emergency procedures, including first-aid facilities.

Keeping a Check

- Assess the degree of contact needed.
- Check that contractors are working to agreed safety standards.
- Encourage contractors to report incidents, near misses and injuries.
- Check for changes of circumstance, such as change of personnel.

MORE...

You can find further information on contractor management in the HSE publication *Using contractors – A brief guide*, INDG368, that you can download from:

www.hse.gov.uk

Reviewing the Work

- Review the job and contractor by examining:

    - How effectively the job was planned and the adequacy of hazard identification and risk assessment.

    - If the work was done in accordance with the method statement.

    - How well the contractor performed and if there were any problems requiring action to be taken.

- Record the lessons learnt for future reference.

Always remember that contractor work can impact on employees and vice versa. The passing on of information regarding work that may affect others is a vital part of safe working with contractors.

Responsibilities for Control of Risk Associated with Contractors and Visitors

At this point, it is worthwhile reviewing the different duties placed on the following:

- **Employers**

    - **Section 3**, the **Health and Safety at Work, etc. Act 1974 (HSWA) –** duty to protect third parties.

    - **MHSWR –** require risk assessments to be carried out and control measures to be implemented.

- **Persons in Charge of Premises**

    - **Section 4**, **HSWA –** to ensure the safety of all persons using the premises and any plant or substance in the premises.

Requirements to Provide Information to Third Parties

The provision of information to third parties relating to hazards and risks is important:

- **Contractors**

    We have already looked at the provision of information to contractors. To recap: **HSWA** and **MHSWR** put a duty on the employer or client to provide sufficient information to the contractor to ensure their safety.

- **Visitors**

    It is usual to give visitors to the workplace written information on emergency procedures, often in the form of a small card or on a visitors' slip. Think about where the visitor is going and what the purpose of their visit is. It may be necessary to supplement the general information with other, more specific, information relating to their particular situation.

- **General Public**

    Information to the general public will include such things as notices and warnings on perimeter fences, gates, etc. Roadworks and other activities that impact on the general public, as well as requiring prominent signage, may be publicised in local newspapers and pre-work notices erected at the site.

Review of Contractor Performance

Once work by a contractor has been completed, the job should be reviewed in order to improve on future contracts.

The review should include:

- Outcomes and achievements of the contractor.
- Adequacy of procedures in place during the work.
- Consideration of any amendments or additions to the procedures that might be needed.
- Recording the overall performance of the contractor and rating it against established criteria.
- Assembling and providing feedback to the contractor.

STUDY QUESTIONS

8. Employers must manage third parties in the workplace. Give **THREE** examples of third parties and briefly outline the key health and safety issues associated with each party.

9. Once work by a contractor has been completed the job should be reviewed in order to improve on future contracts. Outline aspects of the work that should be included in the review.

(Suggested Answers are at the end.)

Consultation with Employees

IN THIS SECTION...

- Consultation with employees contributes to a good safety culture.

- Formal consultation may be required under the **Safety Representatives and Safety Committee Regulations 1977 (SRSCR)** (for organisations with recognised unions) or the **Health and Safety (Consultation with Employees) Regulations 1996 (HSCER)** (for organisations without union representation).

- Informal consultation includes:

  - Discussion groups.

  - Safety circles.

  - Departmental meetings.

  - Employee discussion.

  - E-mail and web-based forums.

- Consultation may be adversely affected by peer group pressures, tokenism and conflicts of interest.

Role and Benefits of Consultation Within the Workplace

Two of the key organisational requirements for developing and maintaining a positive health and safety culture are co-operation and communication (see later) and both of these involve consultation.

In this respect, you should note some of the observations of the Robens Report (which led to **HSWA**):

The involvement of workpeople

"59 We have stressed that the promotion of safety and health at work is first and foremost a matter of efficient management. But it is not a management prerogative. In this context more than most, real progress is impossible without the full co-operation and commitment of all employees."

A statutory requirement to consult

"68 It is generally accepted that there is no credible way of measuring the value of consultative and participatory arrangements in terms of their direct effect upon day-to-day safety performance. Nevertheless, most of the employers, inspectors, trade unionists and others with whom we discussed the subject are in no doubt about the importance of bringing workpeople more directly into the actual work of self-inspection and self-regulation by the individual firm."

"70 We recommend, therefore, that there should be a statutory duty on every employer to consult with his employees or their representatives at the workplace on measures for promoting safety and health at work and to provide arrangements for the participation of employees in the development of such measures."

Source: Health and safety at work, *Report of the Committee 1970-72*, HMSO

Consultation has a direct effect on safety performance

The key **benefits** from consultation are:

* Better employment relations between workers and employers.

* Workers feel more involved and are more likely to co-operate with their employer.

* It creates a safer and less stressful environment, which contributes to a good safety culture.

Formal Consultation

The two sets of regulations concerned with consultation principles are the:

* **Safety Representatives and Safety Committees Regulations 1977 (SRSCR)**; and the

* **Health and Safety (Consultation with Employees) Regulations 1996 (HSCER)**.

Trade Union Appointed Safety Representatives

Under **Section 2(4)** of **HSWA**, safety representatives may be appointed, under the **SRSCR** made by the Secretary of State, by recognised trade unions. These Regulations are accompanied by an Approved Code of Practice and Guidance Notes (L146).

The representatives are chosen from the employees. They are usually selected from persons who have at least two years' experience with their employer or in similar employment, but this is not mandatory. The employer must give the representative time off with pay for the purpose of carrying out their functions as a safety representative, and for training.

A duty lies on the employer under **Section 2(6)** of **HSWA** to consult the representative(s):

"with a view to the making and maintenance of arrangements which will enable him and his employees to co-operate effectively in promoting and developing measures to ensure the health and safety at work of the employees, and in checking the effectiveness of such measures."

Note that this requirement is not optional; the duty is an **absolute** one.

Safety representatives have 'functions' rather than 'duties'. This means they cannot be prosecuted for not specifically complying with a function. So, apart from the general duty placed on them as an employee, no safety representative is legally responsible for accepting (or not objecting to) the course of action taken by their employer, and representatives are not in any danger of criminal proceedings being taken against them should they not carry out any of their functions. However, note that such protection is only afforded to a representative while acting within their jurisdiction.

TOPIC FOCUS

Safety Representatives – Functions and Rights

- **Functions**

  The main function of a safety representative is to represent the employees in consultations with the employer. Other functions include:

  – Investigate potential hazards and dangerous occurrences and examine the causes of accidents at the workplace.

  – Investigate health, safety or welfare complaints by an employee they represents.

  – Make representations to the employer on matters arising out of the above.

  – Carry out inspections.

  – Represent employees in consultations with HSE inspectors.

  – Receive information from inspectors.

  – Attend safety committee meetings.

- **Rights/Entitlements**

  These include:

  – Time off with pay to carry out functions.

  – Time off with pay for necessary training to carry out functions.

  – To be consulted in good time by the employer on:

    – Introduction of measures that would substantially affect the health and safety of employees.

    – Arrangements for getting a competent person to help the employer comply with health and safety requirements.

    – Information to be given to employees on workplace risks and preventive measures.

    – Planning and organising of health and safety training.

    – Health and safety consequences of new technology planned to be brought into the workplace.

  – Access to documents and other information (see later).

Note:

Representatives do not have powers to stop either work or machinery; they may only advise on such matters.

A safety representative may be appointed only by a recognised, independent trade union if they are to receive the legal rights given under the **SRSCR**. To be an **independent** trade union, it must be on the list held by the certification officer and have applied for, and received, the Certificate of Independence from them.

Carrying Out Inspections

A safety representative is entitled to inspect the workplace, or part of it, on **three** occasions:

- If they have not inspected it within the previous three months.

- Where there has been substantial change in the conditions of work.

- After a notifiable accident, dangerous occurrence or notifiable illness, as specified in the **Reporting of Injuries, Diseases and Dangerous Occurrences Regulations 2013 (RIDDOR)**.

The safety representatives should notify the employer of their intention to carry out an inspection, where it is reasonably practicable. The employer shall provide such facilities and assistance, including facilities for independent investigation and private discussion, as the safety representatives may reasonably require.

Entitlement to Information

On reasonable notice being given, the employer must allow the safety representatives to inspect and take copies of any document relevant to the workplace or to the employees whom they represent and which the employer is required to keep by virtue of any relevant statutory provision. (Relevant statutory provisions are listed in the Schedules to **HSWA**.)

Safety representatives are entitled to receive information, under **Regulation 7(2)** of **SRSCR**, from employers. The employer must make available information, within their knowledge, which is necessary to enable the safety representative to perform their function. However, the employer need not disclose information that:

- Is against the interests of national security.
- Would contravene a prohibition imposed by, or under, an enactment.
- Relates specifically to an individual, unless the individual consents to its disclosure.
- Other than for its effects on health and safety, would cause substantial injury to the undertaking.
- Has been obtained by the employer for the purposes of bringing, or defending, any legal proceedings.

Safety representatives are also entitled to receive information from inspectors under **Section 28(8)** of **HSWA**. However, the inspector must not give information which they judge to be irrelevant to the health and safety of the employees. In addition, the inspector must give a copy to the employer of any information given to the safety representative.

This is, of course, only a brief outline of the Regulations, but it is sufficient for examination purposes.

Enforcement of Rights

A safety representative may complain to an Employment Tribunal that the employer has:

- Not allowed them time off for the purpose of carrying out their functions or receiving training.
- Failed to pay them for their time off.

The entitlement for time off for trade union duties and activities is detailed in the **Trade Union and Labour Relations (Consolidation) Act 1992** in the following sections:

- 168 Time off for carrying out trade union duties.
- 168A Time off for union learning representatives.
- 169 Payment for time off under Section 168.
- 170 Time off for trade union activities.

Safety Committees

A duty is placed on the employer, under **SRSCR**, when so requested in writing by at least two union-appointed safety representatives, to establish a safety committee within three months following the request. Again, consultation with those representatives who made the request shall be made by the employer. Representatives of trade unions will also have to be consulted.

- **Functions**

  The function of the safety committee is identified in **Section 2(7)** of **HSWA** – to keep *"under review the measures taken to ensure the health and safety at work of [the] employees and such other functions as may be prescribed"*. There are no other legal requirements concerning the committee's function.

- **Composition/Membership**

  The composition of the committee is a matter for the employer, although they must have at least one safety representative on it and display a notice listing its membership. Membership of the safety committee should be decided following consultations between representatives of the trade unions and the management. Safety representatives are not appointed by this committee.

  It is essential that a proper balance is achieved in the structure of the safety committee; it should have both management and shop-floor representatives on it. It may be useful to elect *ex officio* members, e.g. the medical officer or one of their staff, the works engineer, the production manager and the safety practitioner.

  > **MORE...**
  >
  > The Approved Code of Practice and Guidance Notes (L146), *Consulting workers on health and safety*, contains useful information on safety committees which you can refer to at:
  >
  > www.hse.gov.uk

  The aim is to keep membership of the committee reasonably compact and to ensure a mechanism exists for the consideration and implementation of recommendations by senior management. Although not having executive power, the committee has a strong advisory role to play in the management and resolution of an organisation's health and safety problems.

The committee should meet on a regular basis, circulate an agenda in advance, and keep proper minutes which record what action is to be taken and by whom.

Formal Consultation Directly with Employees

The **HSCER** extend consultation to non-union representatives of employee safety, i.e. non-union workplaces where the **SRSCR** do not apply.

Consultation

Employers should consult employees:

- directly, and/or

- through employee representatives **elected** by a group of employees.

Where consultation is through such employee representatives, the employer must inform the employees of the names of those representatives, and the group of employees they represent. Employees must also be told when the employer discontinues consultation with those employee representatives.

Discontinuation may occur when:

- The employee representatives have informed the employer that they no longer intend to represent their group of employees in health and safety consultations.

- The employee representatives no longer work in the group of employees they represent.

Consultation with workers

- The period of election has elapsed without the employee representatives being re-elected.

- Employee representatives have become incapacitated from performing the duties required under the **HSCER**.

Employees and their representatives must be informed by the employer if they decide to change from consulting with the employee representatives to consulting with the **employees** directly.

TOPIC FOCUS

Representatives (Non-Union) of Employee Safety – Functions and Rights

- **Functions**

  These include:

  - Making representations to the employer on potential hazards and dangerous occurrences which could affect the employees they represent.

  - Making representations to the employer on general health and safety matters (particularly in relation to the matters on which employers are obliged to consult) which may affect the health and safety of the employees they represent.

  - Representing their group of employees in consultations with enforcing authority inspectors.

- **Rights/Entitlements**

  These include:

  - Time off with pay for functions.

  - Time off with pay for training.

  - To be consulted by the employer on the following:

    - Introduction of any measure substantially affecting the health and safety of the employees concerned.

    - The appointment of persons nominated to provide health and safety assistance, and assist in emergency procedures (as required by **Regulations 7** and **8** of **MHSWR**).

    - Any health and safety information the employer is required to provide to the employees or the safety representatives by or under any relevant statutory provision.

    - The planning and organisation of any health and safety training the employer is required to provide by or under any relevant statutory provision.

    - The health and safety consequences of the introduction (including the planning thereof) of new technologies into the workplace.

- To be provided with information from the employer:

  - As is necessary for full and active participation and carrying out of functions, e.g. on risks, preventive measures, etc.

  - From **RIDDOR** reports (applies only to cases where representatives elected).

Provision of Information

Where consultation is direct, employers must provide all information the employees will require in order to participate fully in the consultations. The same applies to employee representatives, who must be given all necessary information to enable them to perform their functions and participate in consultation. These employee representatives must also be provided with information associated with the records to be kept under **RIDDOR** where the information relates to the workplace of the employees they represent (but not to individual employees).

The employer is not obliged to disclose information that:

- Does not relate to health and safety.

- Is against the interests of national security.

- Would contravene any prohibition imposed under any legislation.

- Relates specifically to an individual (unless that individual has given their consent).

- Would damage the employer's undertaking, or the undertaking of another person where that other person supplied the information.

- Has been obtained by the employer for the purpose of any legal proceedings.

Enforcement

As for trade union appointed safety representatives, enforcement is handled through the enforcement agency and tribunals for such things as refusal to grant time off with pay for training, etc.

Informal Consultation

> **MORE...**
>
> You can download *Consulting employees on health and safety: A brief guide to the law* (INDG232) from:
>
> www.hse.gov.uk
>
> The European Agency for Safety and Health at Work has produced a report entitled *Country report – United Kingdom: Worker participation in the management of occupational safety and health – qualitative evidence* from ESENER-2 which focuses on the representation of workers' interests in health and safety as experienced by representatives, fellow workers and employers and managers.
>
> The report is available at:
>
> https://osha.europa.eu/en/publications/country-report-united-kingdom-worker-participation-management-occupational-safety-and/view

Having looked at the formal processes of consultation, you might think that there is no need for informal consultation. Yet, when you look at formal and informal organisations, in many ways the informal route is often more effective in getting things done. Note that the **HSCER** allow employers to consult employees 'directly' without the need to go through union or employee representative channels but give little indication as to how this should be done. Direct consultation would involve a certain amount of bureaucratic procedure to ensure communication and feedback with every member of staff and would still therefore be considered a formal arrangement. How, then, does informal consultation take place?

Opportunities for personal contact occur almost daily in the various meetings which take place between management and employee – workplace inspections, toolbox talks, induction training, safety audits, even staff appraisals. Individuals will often express genuine, personal feelings in a one-to-one situation when free from peer group pressure, in a more open manner than in a group.

- **Discussion Groups**

    These consist of a group of individuals coming together to discuss issues of mutual interest. In the workplace, groups may be formed, often from volunteers, to deal with a number of issues both work and non-work-related. They may be given certain remits, such as safety and quality.

- **Safety Circles**

  These are small groups of employees – not safety representatives or members of safety committees – who meet informally to discuss safety problems in their immediate working environment. The idea is based on the 'quality circles' concept and allows the sharing of ideas and the suggestion of solutions. Any insurmountable problem would be referred to the safety representative or safety committee.

- **Departmental Meetings**

  These meetings are normally attended by shop-floor representatives, supervisory and management staff who meet frequently, often once a week, to discuss general matters affecting their department, such as shift patterns, maintenance and breakdown procedures, and production targets. It is difficult to discuss any of these without impinging on health and safety requirements and, although perhaps not a major objective of such meetings, health and safety policies and arrangements would come under examination. Any health and safety problems identified would probably be referred to senior management through the safety representative or safety committee.

  Informal consultation can take the form of departmental meetings

- **Employee Discussions**

  These are discussions, formal or otherwise, by groups of employees.

- **E-Mail and Web-Based Forums**

  The informal consultation methods outlined above involve face-to-face communication. Talking to other team members, seeking support and guidance from colleagues and discussing work practices and other issues with managers are all examples of this. However, the use of the internet and intranet to get updates and information, and social media options to stay in touch and exchange views, are becoming increasingly important.

  To ensure that communication is effective in the workplace it is important to use all the available and established channels to provide ideas and concerns about health and safety.

  Workers can use:

  - The intranet to access and seek information.

  - E-mails to clarify and engage in two-way communication on health and safety issues.

  - Web-based forums for exchange of information, discussion and debate about contentious issues.

  Electronic methods are also valuable where there are barriers to communication arising from working in remote and isolated work locations associated with:

  - Shift work.

  - Lone working.

  - Isolation by distance or from team support.

  - Working from home.

Behavioural Aspects Associated with Consultation

In any social group, conflict may arise between two or more people, interest groups, genders, ethnic or racial groups, etc. – workplaces are no exception. Safety committee member 'A' serves on the committee to represent their department or perhaps a particular group of workers with common skills. Similarly, committee member 'B' represents **their** department members. A and B, although sharing a common membership of the safety committee, may well be pursuing different objectives. They may both be seeking improved health and safety arrangements for their members but may be in competition for the allocation of limited resources to their particular project.

- **Peer Group Pressures**

  A safety representative serving on a safety committee may feel that they have to question and criticise any suggestion put forward by a management representative on the committee. Remember that the safety representative is a worker's representative; they are neither part of the management team nor necessarily 'a competent person'. Their perception of health and safety problems may be different from that of management and not constrained by budgeting considerations. Their role is mainly a policing one in which they monitor the safety performance of management and, because of peer group pressure, they may see themselves in a conflicting, rather than co-operative, role.

- **Danger of Tokenism**

  One of the dangers associated with consultation is tokenism – where management go through the consultation process but the views expressed by employees are apparently ignored. Clearly, during the consultation process, there is no obligation on the employer to make changes suggested by employees (unless there is a legal requirement) and this may be for perfectly legitimate reasons. However, the employer should respond to information gained during the consultation process and explain what action will be taken and why some proposals may not be implemented, otherwise there may be resentment and apathy toward the process.

- **Potential Areas of Conflict**

  The safety representative may sometimes view themself as an expert on health and safety matters. Conflict may arise between the safety representative and the first line supervisor where the safety representative may have advised their members (wrongly) not to carry out a particular management instruction. This is not to say that conflict always arises as a result of worker attitude toward management. The converse is equally true, with management taking the view that their opinions are correct simply because they are management and think they know better. Consultation about problems where the views of all the participants are considered should lead to effective decisions.

Role of the Health and Safety Practitioner in the Consultative Process

The term 'safety professional' covers such diverse staff as: safety advisers, occupational hygienists, doctors, nurses, safety managers, human resources managers, training officers, facilities managers, ergonomists, engineers and radiation protection advisers. The qualifications range from the highly qualified doctor to the human resources manager who has completed perhaps a non-examination, three-day, basic health and safety awareness course. The health and safety practitioner needs to be a person with a wide range of abilities and a recognised safety qualification at diploma or degree level with Institution of Occupational Safety and Health (IOSH) membership. In relation to the health and safety consultative process, health and safety practitioners have a substantial role to play. They are often the first contact for the employer or worker on health and safety matters. The safety practitioner maintains a number of relationships:

- **Within the Organisation**

  - The position of health and safety practitioners in the organisation is such that they support the provision of authoritative and independent advice.

- The post-holder has a direct reporting line to directors on matters of policy and authority to stop work which is being carried out in contravention of agreed standards and which puts people at risk of injury.

- Health and safety practitioners have responsibility for professional standards and systems and, on a large site, or in a group of companies, may also have line management responsibility for junior health and safety professionals.

- **Outside the Organisation**

Health and safety practitioners liaise with a wide range of outside bodies and individuals, including: local government enforcement agencies; architects and consultants, etc.; the fire department; contractors; insurance companies; clients and customers; the public; equipment suppliers; the media; the police; medical practitioners; and hospital staff.

This is a very wide brief and indicates that the safety practitioner requires a broad and extensive knowledge of health and safety matters in order to fulfil their duties. They are the organisation's first contact when health and safety problems are encountered, and will give advice on short-term safety solutions to problems and follow this through with perhaps a recommendation for a change in policy or the introduction of new technology or new/revised safe systems of work. They will also recommend the services of outside expert consultants where the problem requires scientific, medical or technical advice which is outside their area of expertise. They may also be involved in safety committees in a chairing role or simply in an advisory capacity during committee deliberations.

STUDY QUESTIONS

10. Outline the functions of a safety representative as appointed under the **Safety Representatives and Safety Committees Regulations 1977**.

11. On what matters should an employer consult with employees in a non-union workplace?

12. What is a safety circle?

(Suggested Answers are at the end.)

Health and Safety Culture and Climate

IN THIS SECTION...

- Health and safety culture may be defined as 'a system of shared values and beliefs about the importance of health and safety in the workplace'.

- The health and safety climate is an assessment of people's attitudes and perceptions at a given time.

- Organisational factors, e.g. training, availability of suitable equipment, behaviour of managers, etc. influence individual behaviour.

- There are many indicators of the health and safety culture of an organisation, e.g. housekeeping and relationships between managers and workers.

- Health and safety culture and climate may be assessed by:

    - Perception surveys.

    - Findings of incident investigations.

    - Effectiveness of communication.

    - Evidence of commitment by personnel at all levels.

Meaning of Culture and Climate

There are numerous definitions for a health and safety culture but, essentially, it involves a system of shared beliefs about the importance of health and safety in the workplace.

The definition by the former Health and Safety Commission's Advisory Committee on the Safety of Nuclear Installations is:

"The safety culture of an organisation is the product of individual and group values, attitudes, perceptions, competencies, and patterns of behaviour that determine the commitment to, and the style and proficiency of, an organisation's health and safety management."

"Organisations with a positive safety culture are characterised by communications founded on mutual trust, by shared perceptions of the importance of safety and by confidence in the efficacy of preventive measures."

Source: *ACSNI Human Factors Study Group: Third report – Organising for safety*, HSE Books, 1993

What, then, is **health and safety climate**?

Unfortunately, there is no universal definition and many authors use the terms culture and climate interchangeably. One commonly accepted explanation is given by Professor Sir Cary Cooper who distinguished between three related aspects of culture:

- Psychological aspects – how people feel, their attitudes and perceptions – safety climate.

- Behavioural aspects – what people do.

- Situational aspects – what the organisation has – policies, procedures, etc.

It is generally accepted that safety climate refers to the psychological aspects of health and safety and is measured through a safety climate or attitude survey (see later).

The important thing to remember about a safety culture is that it can be positive or negative. A company with a negative or poor safety culture will struggle to improve safety or prevent accidents, even if they have excellent written procedures and policies and state-of-the-art safety equipment. The reason for this really comes down to people, their attitudes to safety and how this attitude is encouraged and developed.

Influence of Health and Safety Culture on Behaviour and the Effect of Peer Group Pressure and Norms

The safety culture of an organisation has a profound impact on the behaviour of those who work within it. A poor safety culture will tolerate indifferent and even dangerous behaviour which will inevitably become the norm so that even workers well aware of unsafe practices will tolerate poor practices. One such influence is peer pressure from work colleagues.

Group Formation

In a social situation we group ourselves with those of a similar outlook; in the work situation we have little choice as to who we work with. A lot of work situations involve group work or committees and discussion groups. Social groups are an essential part of life, since many activities cannot be performed alone.

Group Reaction

In large groups, the majority scarcely speak at all; there is often a wide variety of personalities and talent. There are differences in behaviour and opinions, discussion is restrained, and disagreement is easily expressed. The group tends to create rules and arranges for division of labour.

Most people prefer to belong to a fairly small group. Each individual can then exert influence on the group, and speak when they wish to, but there is adequate variety of personality to tackle common tasks, and for social purposes.

Social groups are an essential part of life

Group Development

Groups develop an order in terms of the amount of speech and influence permitted. Dominant individuals struggle for status and an order develops, which might not be the one that management would want. Low status members talk little, speak politely to senior members, and little notice is taken of what they have to say.

A person's position in the group depends on their usefulness. The system is maintained. A person who talks too much is stopped. High status members are encouraged to contribute. Group interaction depends on the person's status within the group. When away from the group, a person reverts to their own individual personality. A person can be dominant at work, and yet introverted when away from the workplace.

Group Control

A group will:

- Establish standards of acceptable behaviour or group 'norms'.
- Detect deviations from this standard.
- Have power to demand conformity.

Modifying deviant behaviour involves:

- Ignoring people.

- Verbal hostility and criticism.

- Ridicule.

- Spreading unflattering gossip.

- Harassment.

- Disruption of work.

- Overt intimidation.

- Physical violence.

There are differences between the methods used by groups of men and groups of women. Groups may engage in gender and racial discrimination, which is difficult for management to control.

A lot of safety and health activity tends to be aimed at the individual, when in fact it is much better to target the group. If the dominant leader of the group is very safety-conscious, then safety can quickly become a group norm.

Impact of Organisational Cultural Factors on Individual Behaviour

We are all influenced, to some degree, by things that we see and hear. Billions of pounds are spent on television advertising because companies know how influential television can be – our behaviour is being moulded by an influential medium. In the workplace, who and what are likely to influence our behaviour when it comes to safety?

Typical answers might include:

- **Managers and Supervisors**

  If they appear to condone poor behaviour, then it is likely to go unchecked. Does safe behaviour rank way below productivity? Do they show commitment to safety and lead by example? Do they commit sufficient resources to health and safety?

- **Work Colleagues**

  The way that colleagues behave will probably have an influence on others. What is their attitude to risk-taking?

- **Training**

  Not being trained in correct procedures and use of equipment can affect health and safety. Does the organisation see training as a priority? Is the training appropriate?

- **Job Design**

  Job design may be done in a way that makes safe behaviour difficult. How much consideration has been given to the layout of the job and the needs of the individual?

- **Work Equipment**

  If this is not kept in good order or is unavailable, it may affect health and safety. What is the organisation's attitude to equipment maintenance?

The HSE publication, *Reducing error and influencing behaviour* (HSG48), identifies certain factors associated with good safety performance:

- **Effective communication** – between, and within, levels of the organisation, and comprehensive formal and informal communication.

- **Learning organisation** – the organisation continually improves its own methods and learns from mistakes.

- **Health and safety focus** – a strong focus by everyone in the organisation on health and safety.

All staff are focused on health and safety

- **Committed resources** – time, money and staff devoted to health and safety showing strong evidence of commitment.

- **Participation** – staff at different levels in the organisation identify hazards, suggest control measures, provide feedback and feel that they 'own' safety procedures.

- **Management visibility** – senior managers show commitment and are visible 'on the shop floor'.

- **Balance of productivity and safety** – the need for production is properly balanced against health and safety so that the latter is not ignored.

- **High-quality training** – training is properly managed, the content is well chosen and the quality is high. Counting the hours spent on training is not enough.

- **Job satisfaction** – confidence, trust and recognition of good safety performance.

- **Workforce composition** – a significant proportion of older, more experienced and socially stable workers. This group tend to have fewer accidents, and lower absenteeism and turnover.

Indicators of Culture

TOPIC FOCUS

Indicators of safety culture within an organisation include:

- Housekeeping.
- The presence of warning notices throughout the premises.
- The wearing of PPE.
- Quality of risk assessments.
- Good or bad staff relationships.
- Accident/ill-health statistics.
- Statements made by employees, e.g. 'My manager does not care' (negative culture).

Some of these indicators will be easily noticed by a visitor and help to create an initial impression of the company.

Correlation Between Health and Safety Culture/Climate and Health and Safety Performance

It is quite easy to identify a correlation between cultural indicators and health and safety performance. An experienced safety practitioner can often gauge the standard of safety performance of an organisation from an initial walk-round and first impressions. The standard will often be confirmed on completion of a detailed audit/inspection.

Subjective and Objective Nature of Culture and Climate

The term 'safety culture' can be used to refer to the behavioural aspects (i.e. 'what people do'), and the situational aspects of the company (i.e. 'what the organisation has'). The term 'safety climate' should be used to refer to psychological characteristics of employees (i.e. 'how people feel'), corresponding to the values, attitudes, and perceptions of employees with regard to safety within an organisation.

Safety culture and safety climate are distinct though related concepts. Culture reflects deeper values and assumptions while climate refers to shared perceptions among a relatively homogeneous group. As the two concepts are often used together it is important to recognise that most efforts at measurement, typically through workplace surveys, are assessing climate. Safety climate data can tell us something about the underlying culture, particularly where gaps in perceptions exist within an organisation.

Measurement of the Culture and Climate

MORE...

Loughborough University has developed a Safety Climate Assessment Toolkit which may be downloaded from:

www.lboro.ac.uk/media/wwwlboroacuk/content/sbe/downloads/Offshore%20Safety%20Climate%20Assessment.pdf

While there are many indicators that can give a first impression of a company's safety culture/climate, it is possible to measure some of the indicators to obtain a more accurate picture of the sense of culture within an organisation.

There are a number of measurement tools available.

Safety Climate Assessment Tools

The Health and Safety Laboratory has published a safety climate tool that uses eight key factors mapped around 40 statements on which respondents are asked to express their attitude:

- Organisational commitment.
- Health and safety behaviours.
- Health and safety trust.
- Usability of procedures.
- Engagement in health and safety.
- Peer-group attitude.
- Resources for health and safety.
- Accidents and near-miss reporting.

The kit is available in a software format and will analyse and present the results as charts that can be easily communicated to the workforce.

Perception or Attitude Surveys

These are survey questionnaires (often within a safety climate tool) containing statements which require responses indicating agreement or disagreement. Respondents are asked to indicate to what extent they agree or disagree with each statement, generally using a five-point scale that can then be coded to give a score. High scores represent agreement and low scores disagreement.

It is not difficult to produce a questionnaire about general health and safety which would give some idea as to the safety culture within an organisation. The questionnaire must be worded to avoid bias, and to obtain truthful answers confidentiality is necessary. When carried out properly, these questionnaires can identify underlying anxieties and problems which would be difficult to identify by any other means. Take care, however, to make sure that the questionnaires themselves do not create anxiety or suspicion in the minds of employees. When carried out regularly, attitude surveys can identify trends and it is then possible to quantify how attitudes are changing.

Findings of Incident Investigations

Sometimes during an accident/incident investigation, the underlying cause is identified as 'lack of care'. This may indicate individual carelessness or, where carelessness is found to be the widespread cause of accidents/incidents, then this may be an indicator of poor safety culture.

Where the same underlying cause keeps recurring, the safety manager has to introduce a process of education or re-education of the workforce to encourage a change of attitude. The findings and lessons learnt from incident investigation are invaluable in preventing similar occurrences, setting policy, formulating safe systems of work, writing training materials and, after publication to the workforce, demonstrating company commitment to the principles of good safety management.

Effectiveness of Communication

> **DEFINITION**
>
> **COMMUNICATION**
>
> It is the transfer of information from one person to another with the information being understood by both the sender and receiver.

The process of communication requires a sender, a receiver and feedback. Feedback is the part that is often left out of the process and this is what leads to problems. Successful communication is measured by feedback which allows the sender to test whether the receiver has fully understood the communicated message.

Communication methods are written, verbal or visual, or a combination of all three. The method chosen must be appropriate to the type of information to be communicated and its objectives, the sophistication of the audience (receivers), and the structure and culture of the organisation.

Communication surveys can be used to find out how effectively information has been transferred to new members of staff. A sample of comparatively new members of staff can be interviewed to identify how well they have assimilated the company's safety culture or how much they have retained from company health and safety training. This type of survey can be done formally or informally.

Effective communication involves:

- Including everyone who should be included.

- Not overloading people with large quantities of information; prioritise anything urgent.

- Being brief, direct and keeping it simple.

- Being fast but not at the expense of accuracy.

- Being selective; sending only what is necessary.

- Encouraging feedback to ensure the message has been received and understood.

- Using as few links in the communication chain as possible to prevent distortion of the original message.

Evidence of Commitment by Personnel at all Levels

DEFINITION

COMMITMENT

A declared attachment to a doctrine or cause.

It is the goal of the health and safety practitioner to ensure commitment to health and safety by everyone within an organisation. This commitment must start at the management board level. It is essential that management show their commitment to safety as this sets the standard for the whole organisation. The workforce will only believe in this commitment if they know that management are willing to sacrifice productivity or time in order to ensure worker safety.

Evidence of commitment can be seen by management visibility. If managers are not seen on the 'shop floor' or at the 'sharp end of activity', workers may assume that they are not interested in the job or health and safety. Lack of management visibility is seen as a lack of commitment to safety and this becomes part of the organisation's safety culture.

Visible commitment can be demonstrated by management:

- Being seen and involved with the work and correcting deficiencies.

- Providing resources to carry out jobs safely (enough people, time and money, providing appropriate PPE, etc.).

- Ensuring that all personnel are competent (providing training and supervision).

- Enforcing the company safety rules and complying with them personally (introducing safe systems of work and insisting on their observance).

- Matching their actions to their words (correcting defects as soon as is reasonably practicable, avoidance of double standards).

STUDY QUESTIONS

13. Define the term 'safety culture'.

14. How may the safety climate of an organisation be assessed?

15. Name three ways in which management commitment can be demonstrated.

(Suggested Answers are at the end.)

Factors Affecting Health and Safety Culture and Climate

IN THIS SECTION...

- A positive health and safety culture is promoted by a combination of management commitment, positive leadership, effective training and the setting (and meeting) of targets. It requires a high business profile to be given to health and safety, with involvement, consultation and the promotion of ownership.

- A negative health and safety culture is fostered by organisational change, a lack of confidence in organisational objectives and methods, and inconsistent signals from management.

- Effective cultural change needs good planning and communication, strong employee engagement, training and performance measurement, strong leadership and appropriate feedback.

- Problems with culture change may arise from attempting to change too quickly, a lack of trust in communications or resistance from those not committed to change.

Promoting a Positive Health and Safety Culture or Climate

Management Commitment and Leadership

The most important thing is 'leading by example'. As soon as management undermines the safety standards in order to increase productivity, or ignores an unsafe act, then they lose employee respect and trust and the whole safety culture of the organisation is threatened. It is important to ensure that management behaviour is positive in order to produce positive results and a positive culture.

High Business Profile to Health and Safety

A positive health and safety culture can be promoted by including safety in all business documents and meetings. All newsletters, minutes of meetings, notices, advertisements and brochures can include an appropriate reference to safety; it could simply be reference to the organisation's commitment to safety (e.g. a safety phrase appearing on all notepaper) or, with respect to meetings, it could be an opportunity for any safety concerns to be raised. If safety is seen as an integral part of the business then the profile of safety will be raised.

Health and safety law poster

Provision of Information

It is really important to provide information about health and safety matters in the form of posters, leaflets or in staff newsletters.

Involvement and Consultation

It is vital to involve staff members in health and safety matters. Areas in which staff representatives or health and safety representatives can be actively involved include:

- Risk assessments.

- Workplace inspections.

- Accident investigations.
  - Safety committee meetings.

It is also a legal requirement to consult with employees in good time regarding:

- The introduction of any measures that may substantially affect their health and safety.
- The arrangements for appointing or nominating competent persons.
- Any health and safety information to be provided to employees.
- The planning and organisation of any health and safety training.
- Health and safety consequences of introducing new technology.

Involving and consulting with employees is an important process for getting employees to take ownership of health and safety issues. The fact that they or their colleagues have been involved in health and safety matters encourages respect for safety rules and improves attitudes towards safety. These values all help to produce a more positive safety culture within the organisation.

Training

Training is vital to ensure that people have the right skills to carry out their job safely. Training also makes individuals feel valued and is an important part of their personal growth and achievement. Employees who receive training are more likely to be motivated and take newly learnt skills or ideas back to the workplace.

Promotion of Ownership

There are many ways to promote ownership in individuals. We have mentioned involvement and consultation already, but simply talking to people and asking their opinion or their thoughts on a health and safety problem can encourage them to think about health and safety and what they can do to improve it.

Setting and Meeting Targets

Setting safety targets for individuals or teams can have a positive effect on a safety culture. Usually, there will be an incentive, perhaps a bonus, linked to performance-related pay or an award or prize. The target could be, for example, to obtain a higher score in a health and safety inspection.

Aiming for the target should encourage people to work together in order to achieve it and this usually means people talking about health and safety and ways to improve it.

Once the target is met, that standard must be maintained and further improvements encouraged by setting another target. The targets must, however, be achievable in order to prevent employees becoming disheartened and abandoning the target.

Factors That May Promote a Negative Health and Safety Culture or Climate

There are a number of factors that may contribute to a negative health and safety culture or climate.

Organisational Change

Company reorganisations often leave individuals worried about job security and their position in the organisation. Many people fear change and, unless it is handled correctly, will mistrust management and become suspicious of any alterations to their role or environment (even ones that are beneficial).

Reasons for company reorganisation may be:

- A merger.

- Relocation of the business.

- Redundancies.

- Downsizing.

- External pressures over which the organisation has no power.

Companies may offer voluntary redundancies to make the job losses more acceptable but sometimes the redundancies are compulsory. The company may also offer generous financial packages in excess of the statutory minimum to soften the blow to employees. Problems may occur, however, when the retained staff have to work with reduced manpower and resources. The remaining employees may feel threatened by the possibility of further redundancies, leading to bitterness and anger. Further resentment may develop where shareholders and directors are seen to benefit from the loss of colleagues who have left the business.

Where outside pressures are the cause of the reorganisation, employees may be more understanding than if the changes are brought about by the need to improve profits.

Frequent reorganisations can be damaging to a company unless they are handled well. Increased workforce dissatisfaction may lead to some employees leaving, which in turn can leave gaps in the operation which cause further difficulties. This type of situation can lead to more accidents and incidents as well as increased sickness and absence from work.

Lack of Confidence in Organisation's Objectives and Methods

Most companies have objectives relating to productivity and safety. If productivity appears to take precedence over safety, however, then worker perception will be that the company is unethical and untrustworthy with little commitment to safety, which will lead to a subsequent deterioration in the safety culture.

Examples where workers may feel that safety has been compromised in order to achieve productivity include:

- Safety improvements only made after incidents have occurred.

- Double standards in the application of safety regulations by safety advisers and management.

- Unsafe practices ignored in order to improve productivity.

- Permit-to-work systems not being operated as they should be.

Unsafe practices may be ignored
to improve productivity

- Changes made to safety rules during operation.

Uncertainty

Security is a basic human need. In an uncertain environment, people generate feelings of insecurity. When security cannot be assured, humans cannot achieve their full potential. Uncertainty about the future can lead to dissatisfaction, lack of interest in the job and generally poor attitudes toward the company and colleagues.

Uncertainty is often caused by management behaviour which sends mixed behaviour signals to the workforce. If management are seen to say one thing and then do something different, this undermines their authority and credibility, e.g. managers drinking on the job or failing to wear PPE.

Management Decisions That Prejudice Mutual Trust or Lead to Confusion Regarding Commitment

Management decisions that are, or are perceived to be, inconsistent or poorly made can generate unrest and distrust in an organisation. There may be good reasons for the decision which is why it is extremely important that management are aware that good communication is an important part of the decision-making process.

Circumstances that could give rise to distrust and doubt about management commitment generally (these could equally apply to decisions about safety) include:

- Where there are no rules or no precedents, decisions may appear to be arbitrary and inconsistent.

- Employees are expected to wear PPE whereas visitors or managers are not.

- Refusal to delegate decision-making, leading to demotivation and diminution of a sense of responsibility in subordinates.

- Constant rescinding by senior management of decisions made at lower levels of management.

- Delays in making decisions.

- Decisions affected by conflicting goals between management and worker.

- Decisions affected by conflicting goals between different departments.

- Lack of consultation prior to decision-making.

Effecting Change

There are three factors that should be considered when managing a change in culture:

- Dissatisfaction with the existing situation, e.g. too many near misses.

- A vision of the new safety culture.

- Understanding how to achieve it.

Change is an inherent part of modern life but there are many people who find change difficult to deal with and who are afraid of it. In order to effect change within an organisational culture, you have to plan the strategy and communicate from the beginning in order to involve employees and not alienate them.

Planning and Communication

Planning for change should start at the top of the organisation but should encourage participation at all levels. There should be clear objectives as to what is to be achieved by the proposed change, e.g. a cost-benefit analysis of the changes suggested.

Plans for change should clearly designate who is responsible for initiating and implementing specified changes as well as how each stage of the change process will be conducted. Effective communication between all those implementing change is crucial.

To prevent rumours circulating and misunderstandings developing, it is important to publicise information relating to the pending change as early as possible. Wherever possible, direct briefings, meetings or interviews should keep managers and staff aware of proposed changes and the progress made as changes get under way.

Strong Leadership

Managers at all levels need to demonstrate strong leadership and not give inconsistent or mixed messages.

A Gradualist (Step-by-Step) Approach

One of the ways of effecting change in an organisational culture is by taking a gradualist (step-by-step) approach, with changes phased in over a period of time. The main advantage of this approach is that it ensures that there is time for adaptation and modification; it also allows time for the change to become part of the established culture.

The major disadvantage of this approach is that the changes take a relatively long time to implement. This can mean that unsatisfactory conditions and mindsets may be left in place for longer than is desirable.

Action to Promote Change

- **Direct**

  This is where positive action is carried out with the sole objective of effecting change, perhaps by setting up a two-tiered system, i.e. a steering group and a working party. The steering group should consist of high-level personnel (e.g. directors and heads of departments) who give broad objectives, set timescales and meet approximately every three months. The working party, however, will meet every month and will consist of middle management, first-line supervisors and union/worker representatives. The working party will carry the 'message' to the workforce and provide feedback.

  The chair of the working party should also be a member of the steering party and this role is usually filled by a safety professional who can act as the link between the two groups.

  The pace of change should be dictated by the feedback given by the working party.

- **Indirect**

  Indirect methods bring about change but they are not necessarily the primary reason for carrying out the method. For example, risk assessments identify deficiencies in the workplace and corrective action to put them right. Widespread use can indirectly encourage a risk assessment mindset or attitude (a culture of greater awareness of risks, etc.).

Strong Employee Engagement

Cultural change is not the sole responsibility of management; there also has to be significant commitment from employees who must recognise the need for change.

Training and Performance Measurements

- **Training courses** can include information about new or impending safety legislation or safety technology, thereby indirectly paving the way for future changes.

- **Performance measurements** can be introduced to encourage employees to have a greater interest and involvement in health and safety. Where performance measurements improve over time, they can be linked to an incentive scheme, but they should not be linked to accident/incident rates as this can lead to under-reporting. Performance measurements are an inexpensive way of promoting health and safety, but they need the support of management and unions to be successful.

Importance of Feedback

Feedback is crucial to ensure that any changes implemented are working successfully. Feedback from employees will enable management to evaluate the new processes, and fine-tune them where necessary.

Problems and Pitfalls

In many cases, the introduction of change within an organisation is often accompanied by problems such as conflict. Problems associated with change include:

Change is often accompanied by conflict

- **Changing Culture Too Rapidly**

  Where changes have occurred too quickly, employees may feel extremely vulnerable, insecure, confused and angry.

  Where the changes have brought together new personalities, then conflict between individuals may occur. Differences of temperament are at their most obvious when people are new to each other; a measure of tolerance may build up over time.

- **Adopting Too Broad an Approach**

  Trying to do too much all in one go can dilute the resources so that little impact is seen. It is better to target resources on fewer, manageable issues. It is important to be clear about what the objectives are at the start so that everyone is aware of the changes that will occur.

- **Absence of Trust in Communications**

  This is unsettling and demotivating. Inconsistent management behaviour can lead to mistrust and uncertainty causing a complete breakdown in relations between management and the workforce. Poor communications in periods of change can lead to misunderstanding and confusion, which can fuel conflict.

- **Resistance to Change**

  Some people are more resistant to change than others. Older people tend to be more resistant than young people, and people with heavy financial commitments tend to fear change as they need to feel secure.

  Some people develop set patterns of thought and behaviour which can be difficult to overcome when change occurs. This is known as **perceptual set**, and is the way in which observed information is processed by the individual to fit their internal experience, attitude, expectations, sensitivity and culture.

All these factors need to be considered and dealt with as part of the change process.

STUDY QUESTIONS

16. Identify the features of a positive health and safety culture within an organisation.

17. Briefly explain what is needed to effect cultural change within an organisation.

(Suggested Answers are at the end.)

Summary

Types of Safety Leadership

Safety leadership is *"the process of influencing the activities of an individual or a group in efforts toward goal achievement in a given situation".*

Types of safety leadership include:

- Transformational.
- Transactional.
- Servant.
- Situational and contextual (Hersey and Blanchard).

Benefits of Effective Health and Safety Leadership

The **HSE/IOD** have issued guidelines to promote the effective leadership of health and safety.

Effective **health and safety management** involves leaders at all levels in an organisation understanding the range of health and safety risks and recognising their importance.

Effective **safety leadership** can influence the achievement of a positive health and safety culture in an organisation, and the tangible benefits of a positive health and safety culture are reflected in indicators of good health and safety performance.

Employee consultation and involvement is an essential element of effective health and safety management and leaders play an essential role in promoting the participation and engagement of the workforce.

Both the **safety practitioner** and the **organisation** have a role in encouraging effective health and safety leadership in order to achieve high standards of health and safety in the workplace.

The **Financial Reporting Council** has issued guidance as to how companies may assess the effectiveness of their risk.

Internal and External Influences

Internal influences on the organisation include:

- Finance.
- Production targets.
- Trade unions.
- Organisational goals and culture.

External influences include:

- Legislation.
- HSE/Parliament.
- Enforcement agencies.
- Courts/tribunals.
- Contracts/contractors/clients.

- Trade unions.
  - Insurance companies.
- Public opinion.

Types of Organisations

An **organisation** is a group of persons who interact in order to achieve certain predetermined goals or objectives.

In a **formal** organisation, the organisation's structure is based on relationships from the chief executive down. This hierarchical structure is represented by the company organisation chart, or organogram.

The **informal** organisation is represented by individual and group behaviour, and depends on the quality of personal relationships.

The organisation can be viewed as a **system**; different parts of an organisational system are functionally interrelated – change in one part affects other parts of the organisation.

Conflict may arise as a result of individual goals not being consistent with those of the organisation.

Requirements for Managing Third Parties

Definitions in this area:

- Third parties:
  - Contractors.
  - Agency workers.
  - Other employers (shared premises).

When using **contractors**, procedures need to be adopted to ensure:

- Planning – including risk assessment.
- Selection – competent contractor.
- During contract – ensure contractor is inducted and is aware of local procedures.
- Checking of performance.
- Review of procedures.

Consultation with Employees

With regard to **formal consultation**, a duty is placed on employers by **Section 2(6)** of **HSWA** to consult appointed trade union safety representatives.

The functions of **safety representatives**, as laid down in the **SRSCR**, include to:

- Represent the employees in consultation with the employer.
- Carry out inspections of the workplace.
- Look at causes of accidents.
- Receive information from health and safety inspectors.
- Attend safety committee meetings.

Under the **SRSCR**, a duty is placed on the employer, when so requested by at least two safety representatives, to establish a **safety committee**. The **function** of the safety committee is to keep under review the measures taken to ensure the health and safety at work of the employees, and such other functions as may be prescribed.

The **HSCER** allow for consultation with non-trade-union representatives of employee safety. Employers may consult directly with the workforce, or through elected representatives of employee safety.

Informal consultation can be a valuable source of information; it can take place in various ways, e.g. discussion groups, safety circles, departmental meetings.

Consultation may be compromised by peer group pressure, tokenism and areas of conflict.

Health and Safety Culture and Climate

Health and safety culture may be defined as a system of shared values and beliefs about the importance of health and safety in the workplace.

Health and safety climate is an assessment of people's attitudes and perceptions at a given time.

Organisational factors, e.g. training, availability of suitable equipment, behaviour of managers, etc. influence individual behaviour.

There are many **indicators** that give a first impression of a company's health and safety culture. It is also possible to measure indicators that give a more accurate picture; these include:

- Attitude surveys.
- Prompt lists.
- Findings of incident investigations.
- Effectiveness of communication.
- Evidence of commitment by personnel at all levels.

Factors Affecting Health and Safety Culture and Climate

A **positive** health and safety culture can be promoted by various factors, such as: the commitment of management, a high business profile, provision of information, involvement and consultation, training, promotion of ownership and the use of targets.

A **negative** health and safety culture can also be affected by various factors, such as: organisational change, lack of confidence in an organisation's objectives and methods, uncertainty and inconsistent management decisions.

A **change in attitudes** can be achieved by planning and communication, and should be introduced using a gradualist approach. Action to promote such a change can be direct or indirect.

Exam Skills

By now you should be familiar with the style of NEBOSH exam questions; the next one is a straightforward, 10-mark question on health and safety culture.

QUESTION

Outline how safety tours could contribute to improving health and safety performance and to improving health and safety culture within a company. Discussion of the specific health and safety requirements, problems or standards that such tours may address, is **not** required. **(10)**

Suggested Answer Outline

Remember this is a 10-mark question, so try to identify 12 points in order to gain full marks. The examiner would be looking for some of the following points to be included in your answer.

Safety tours can be used in an organisation to help improve its health and safety performance and culture by:

- Identifying compliant and non-compliant behaviours.
- Ensuring compliance with legislation and good practice.
- Seeing how effective its actions are.
- Establishing that new programmes are working as expected.
- Identifying good practice across the company.
- Consolidating good relationships with the workforce during tours.
- Assessing workforce behaviour on an unscheduled basis.
- Spotting local issues.
- Identifying company-wide issues.
- Demonstrating leadership/engagement and commitment.
- Highlighting management commitment.
- Ensuring that local remedial actions to solve issues raised have been implemented.
- Encouraging local ownership of health and safety.
- Highlighting the importance of safety.
- Combining it with other types of tours (quality/environmental, etc.) saving time/resources, etc.
- Sharing the findings with the workforce, showing openness.
- Making it easier to communicate on a regular basis with employees.

Example of How the Question Could Be Answered

Safety tours can contribute to improving health and safety performance and the culture, as they should be used to identify good and poor health and safety behaviours on the shop floor of the organisation. This information can be used in a simple manner to trend on performance against set behaviours via a performance feedback sheet, such as PPE being worn v. not worn – has it improved since last time or got worse?

The tours can be used to target new understanding or compliance with new initiatives and programmes the company introduces and is an effective way of engaging the workforce across an organisation to support these. The actions raised should be solved locally, visually and quickly, which demonstrates leadership and commitment of the company and enables best practice sharing. If the safety tours are carried out by managers, they can be an effective demonstration of management commitment to safety; however, this does require action to be taken as a result of the tour.

Reasons for Poor Marks Achieved by Candidates in Exam

An exam candidate would achieve **poor marks** for:

- Describing how to **carry out** a safety tour.

- Looking at specific issues, although the question particularly said not to.

- Focusing on the timing/frequency of tours and not looking at how tours can help improve health and safety performance.

Element A10

Human Factors

Learning Outcomes

Once you've read this element, you'll understand how to:

1. Outline psychological and sociological factors which may give rise to specific patterns of safe and unsafe behaviour in the working environment.

2. Explain the nature of the perception of risk and its relationship to performance in the workplace.

3. Explain the classification of human failure.

4. Explain appropriate methods of improving individual human reliability in the workplace.

5. Explain how organisational factors can contribute to improving human reliability.

6. Explain how job factors can contribute to improving human reliability.

7. Outline the principles, conditions and typical content of behavioural change programmes designed to improve safe behaviour in the workplace.

Contents

Human Psychology, Sociology and Behaviour

IN THIS SECTION...

- Human behaviour is influenced by the personality, attitude, aptitude and motivation of the individual.
- Different theories have been proposed to help explain what motivates individuals at work.
- Experience, social background and education/training affect behaviour at work.

Meaning of Terms

It is important that you understand what the terms 'psychology' and 'sociology' mean and the impact of these on the area of health and safety.

> ### DEFINITIONS
>
> **PSYCHOLOGY**
>
> A study of the human personality.
>
> **SOCIOLOGY**
>
> A study of the history and nature of human society.

Influence of Personality, Attitude, Aptitude and Motivation on Human Behaviour

- **Personality** – the combination of characteristics or qualities that form an individual's distinctive character.

  The main dimensions of personality are:

  - Extroversion/introversion – extroverts are more outgoing than introverts.
  - Neuroticism – neurotics have high levels of anxiety.
  - Conscientiousness – such people tend to be well organised.
  - Agreeableness – these people are more willing to co-operate with others and avoid conflict.
  - Openness to experience – such people tend to welcome new experiences and are more curious.

  Some of these characteristics, such as conscientiousness and agreeableness, are likely to be positive traits in terms of health and safety.

- **Attitude** – reflects how a person thinks or believes about something (often called the object of the attitude) and this may then extend to how they behave. For example, a person involved in a workplace transport accident is more likely to become safety conscious (at least initially) in relation to work transport hazards as a result.

- **Aptitude** – the ability of an individual to undertake a given task safely. Training and supervision usually increase aptitude.

- **Motivation** – the factors that influence an individual to behave in a certain way. Most people are generally motivated to avoid accidents and ill health, although other motivators may conflict with the general principle. For example, wearing Personal Protective Equipment (PPE) may be uncomfortable and interfere with the task and so may encourage people to take greater risk by not using it.

Key Theories of Human Motivation

There are a number of theories that have been developed to try to explain why people do what they do.

Mayo (Hawthorne Experiments)

One of the most significant contributions to the study of work groups took place in 1927 at the Hawthorne Works of the Western Electric Company in the USA. The Hawthorne Experiments were originally designed as a short project to study the 'relation of quantity and quality of illumination to efficiency in industry'.

What surprised the observers, who had thought they could predict what was going to happen, was that output varied with no relationship to the amount of illumination.

The observers then realised that motivation of the individual is not just to do with money or intensity of lighting. They started to ask themselves what motivated individuals working in groups.

A psychologist named **Elton Mayo** was allowed to conduct a series of experiments, and important findings included:

- Working in a small, harmonious group can have a significant effect on productivity.

- Having a chance to air grievances seems to be beneficial to working relationships.

One of the essential principles here is that when you show concern for or pay attention to people, it can spur them on to perform better. In other words, just the fact that the workers were being studied improved their performance. This is known as 'the Hawthorne Effect' or the 'somebody upstairs cares' syndrome.

Maslow (Hierarchy of Needs)

Abraham Maslow suggested five levels of human need, which they arranged in a hierarchy.

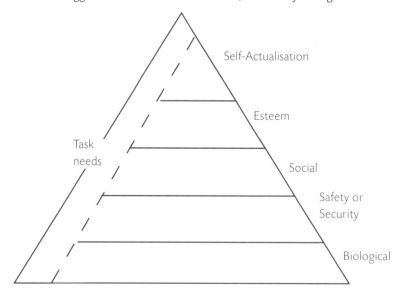

The order in which the needs are listed is significant in two ways:

- It is the order in which they are said to appear in the normal development of the person.

- It is the order in which they have to be satisfied and if earlier needs are not satisfied, the person may never get around to doing much about the later ones.

From this theory, you might expect that people in a poor society will be mostly concerned with physiological and safety needs, whereas those in an affluent society will manage to satisfy those lower needs in the hierarchy and, in many cases, will be preoccupied with the need for self-actualisation. However, Maslow's hierarchy stresses that co-operation can occur only at higher levels between mature individuals, the lower-order needs leading to conflict between individuals. Yet, primitive tribes seem to co-operate more than advanced societies, where conflict between individuals is encouraged. This seems to suggest that there may be a flaw in Maslow's analysis.

The need for self-actualisation refers to the person's need to develop their full potentialities; the meaning varies from person to person, for each has different potentialities. For some, it means achievement in literary or scientific fields; for others, leadership in politics or the community; for still others, merely living their own lives fully without being unduly restrained by social conventions. 'Self-actualisers' are found among professors, businessmen, political leaders, artists and housepersons.

Not all individuals in any one category are able to achieve self-actualisation; many have numerous unsatisfied needs and, because their achievements are merely compensations, they are left frustrated and unhappy in other respects.

Vroom (Expectancy Theory)

V H Vroom defined motivation as a process whereby the individual makes choices between alternative forms of voluntary activities. Employee effort leads to performance and performance leads to rewards; so the choices made by the individual are based on estimates of how well the **expected** results of a given **behaviour** will lead to the **desired** results.

In Vroom's analysis, motivation is based on three factors:

- The **expectancy** that effort will lead to the intended performance, i.e. the individual's confidence in what they are capable of doing and that it will lead to the required outcome. This depends on factors such as resources, skills and support.

- The **instrumentality** of this performance in achieving a particular result, i.e. the perception of individuals of whether they will actually receive what they desire.

- The **desirability** of the result to the individual (**valence**), i.e. whether the person values the outcome.

If employees are going to be motivated then all three factors must be positive and if any are not achieved, employees will not be motivated. Perceptions are key to this theory so even if an employer thinks they have provided everything appropriate for motivation, it is still possible that some individuals will not see it this way.

Expectancy theory can help managers understand how individuals are motivated to choose among various behavioural alternatives. Managers may need to use systems that tie rewards very closely to performance in order to enhance the connection between performance and outcomes. Managers also need to ensure that the rewards provided are deserved and wanted by the recipients. To improve the connection between effort and performance, managers should use training to improve employee capabilities and help employees believe that added effort will, in fact, lead to better performance, including health and safety performance.

Blanchard

According to **Ken Blanchard**, people have a natural desire to grow, develop, and do meaningful work. The key psychological needs of an individual are **autonomy**, **relatedness** and **competence**. If these are satisfied in a workplace, employees will become highly motivated and more engaged.

The Blanchard model identifies a spectrum with six **motivational outlooks**:

- **Disinterested** – the person finds no value in the project or task and considers it a waste of time.

- **External** – the project or task only provides the person with an opportunity for more money or other external gain.

- **Imposed** – the person participates in the project or task only because they feel pressured to do so.

- **Aligned** – the person links participation to a significant value such as learning from others or having others learn from them.

- **Integrated** – the person participates in the project or task because they can link it to a life or work purpose important to them.

- **Inherent** – the person enjoys the activity and wants to participate.

Note that:

- **Disinterested**, **external**, and **imposed** are termed 'sub-optimal motivational outlooks', and reflect low-quality psychological needs and self-regulation.

- **Aligned**, **integrated**, and **inherent** are termed 'optimal motivational outlooks', and serve to satisfy psychological needs.

Optimal motivation theory can help managers understand how individuals are motivated and consequently how a leader's role might activate such motivation in the workplace. This has relevance in enabling leaders to take a more strategic approach and assumes that one of the roles of a leader is to help individuals explore why they are motivated, uncover the reasons for their current motivational outlook, and then use best practices to help facilitate people's shift to a more optimal motivational outlook.

Factors Affecting Behaviour

Experience

The **Management of Health and Safety at Work Regulations 1999 (MHSWR)** as amended, require risk assessments to consider groups who are particularly at risk. The young and inexperienced are specifically mentioned.

With increasing experience, we would expect employees to become more competent and to increase in ability to cope with situations, but, there may also be complacency and a tendency to cut corners.

Age and experience are correlated with differences in accident susceptibility as the following graph indicates. Though its exact shape will vary with circumstances, the curve will remain roughly the same.

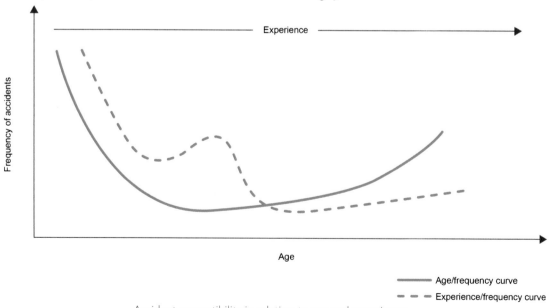

Accident susceptibility in relation to age and experience

Social and Cultural Background

An individual's background will initially have a significant effect on their behaviour. The society and culture in which an individual is brought up teaches acceptable and unacceptable behaviour. Being brought up in a society where there is little work and it is a struggle to find adequate food is likely to make people undertake tasks with less consideration for safety compared to those from more affluent cultures. This relates to Maslow's hierarchy of needs.

Education and Training

Education is often thought of as the act of acquiring knowledge, whereas training is thought of as giving a person more specific skills for a particular task. In many ways, the concept of safety training is a myth. Certainly, the notion of 'bolt-on' training to accompany job training is, at best, misguided. Training that teaches employees how to perform tasks correctly, teaches them how to perform those tasks safely at the same time. Remember the saying: 'the right way to do the job is the safe way to do the job'.

Safety training should be integrated into skills training

The safety practitioner and the training manager should ensure that safety is built into the training package when identifying training needs. In this way, safety is further integrated into the quality and efficiency programme and not left outside it, where money is not available during lean years and time is not available during boom years. Of course, there are times when safety training has to stand alone, such as during induction training, when new employees should be told of specific safety procedures, e.g. fire procedures, first-aid arrangements, etc. In general, however, the more safety training can be integrated into skills training, the better.

If successful, both education and training will secure a positive change in the behaviour of personnel. **It is therefore essential to identify the changes in behaviour required before training begins** and to set outcomes that can be demonstrated after the training has been completed. This approach allows the success of the training to be measured and evaluation and feedback on success to be provided.

STUDY QUESTIONS

1. Define the terms 'psychology' and 'sociology'.

2. Draw a diagram to explain Maslow's hierarchy of needs.

(Suggested Answers are at the end.)

Perception of Risk

IN THIS SECTION...

- Each of our senses sends signals to the brain.
- We tend to screen out things we are not interested in.
- Sensory defects increase with age and ill health.
- Perceptual set is dangerous because we assume both the danger and the solution without seeing the real issues.
- Our perception of hazards can be distorted.
- Errors of perception can also be caused by physical factors, such as fatigue and stress.

Human Sensory Receptors

The natural senses are:

- Sight.
- Hearing.
- Taste.
- Smell.
- Touch.

(Note that there are others, such as the sense of temperature and the sense of acceleration.)

Personal safety involves reacting to the signals sent by our human sensory receptors to the brain.

For example, the eyes send small electrical signals to the brain, where the visual image is constructed and interpreted. Sometimes the eyes see an image one way and the brain interprets it differently. Consider this example of an optical illusion:

Which centre dot is larger?

They are, in fact, the same size.

Each of the other senses also works by sending signals to the brain. There is a time interval between the signal being sent from the sensory receptor and the brain making the person aware of the situation. Remember that the senses are the main way that people get a warning of danger.

Optical illusion

Sensory Defects and Basic Screening Techniques

You will be aware, on occasions, of not seeing or hearing something that was very plain to someone else. Sensory defects increase with age and failing health and some people may need spectacles and hearing aids to compensate for these deficits. The safety practitioner probably needs to be more concerned about those who don't know that they have sensory defects or who try to forget about them.

People also have the ability to shut out things that they are not interested in, i.e. screen out those things that they consider are not worth concentrating on. For example:

- Someone may live within two miles of a motorway, but really have to concentrate to hear the traffic on it.

- A worker is able to filter out background noises in a workshop and maintain a conversation.

- When driving a car, or a work machine, or typing at a keyboard, most of the operations are done in 'auto-pilot' mode. This saves effort and allows us to concentrate on other things, or think ahead.

Screening is a useful asset to have, but it is the reason for many accidents. You cannot expect 100% concentration on safety matters from others if you seldom give 100% attention yourself.

Perception

Perception of Danger

Research into perception of danger by individuals and groups shows that there is a clear distinction between how we perceive risks to personal safety, dangers to health, and dangers to society. Individuals who take part in hazardous sports and activities may be very reluctant to take even a small risk in the work situation.

The factors involved in perception are:

- **Signals from the Sensory Receptors**

  Our eyes, ears, nose, touch and perhaps taste make us aware of the situation, but these signals can be misleading if we suffer from a sensory defect.

- **Expected Information from the Memory**

  We have an expectation of what to see and hear; this signal is from the memory. We sometimes see things which are not there, and don't see things which are. This signal can also be misleading, particularly if it is affected by stress, alcohol, drugs, fatigue or just familiarity.

> **DEFINITION**
>
> **PERCEPTION**
>
> The recognition and interpretation of sensory stimuli, based chiefly on memory.

These two signals combine to give us a 'picture' of the situation of hazard, which is then processed by the brain. We then take, or decide not to take, action.

Perceptual Set

This is sometimes called a 'mindset'. We have a problem – and immediately perceive not only the problem, but also the answer. We then set about solving the problem as we have perceived it. Further evidence may become available which shows that our original perception was faulty, but we are now so pleased with our intelligent solution that we fail to see alternative causes and solutions. **This is a basic cause or factor in many accidents and disasters**.

Students often get such mindsets when answering examination questions and assignments. You have prepared yourself well for a particular type of question. This seems to be there on the examination paper and you immediately set about writing the answer. Later, when discussing this with others or re-reading the question, you wonder how you could have missed the point. The examination committee spent a great deal of effort to make it perfectly clear what was needed, but all to no avail.

The same thing can happen in work situations. For example, a signaller was expected to check that there was a red light at the back of every train which passed his signal box. They had never seen a situation where this was not so in the 10 years that they had been doing his job. However, on one occasion, part of the train had become uncoupled, but they distinctly remembered checking and 'seeing' the red light as the train passed. A following train collided with the part of the train that had become uncoupled. This was a typical case of mindset or perceptual set.

Perceptual Distortion

Everyone's perception of hazard is faulty because it gets distorted. Things that are to our advantage always tend to seem more right than those that are to our disadvantage. Management generally tend to have a different perception of hazard from that of workers. When it affects work rates, physical effort or bonus payments, workers also suffer from perceptual distortion.

Errors in Perception Caused by Physical Stressors

In examining the cause of errors in perception, we also need to consider the effects of fatigue, overwork, overtime, stresses from the workplace, and stresses from home and outside activities. Shift work is a major factor. Our bodies operate best when we have a regular routine. There is an inbuilt clock (the circadian cycle – the internal body clock which dictates when the body should be active and when it should rest), and the pattern of work, rest and sleep is upset by a change of work pattern. It is even possible that we are locked into a seven-day pattern.

Perception is affected by having to keep awake and alert when the body is saying that it is time to sleep. Fatigue is more than tiredness of the muscles and the mind; there is a physical, mental and psychological dimension.

Perception and the Assessment of Risk

Fatigue can lead to errors in perception

If there are problems in our basic perception of a situation, then there are going to be errors in our perception of risk. In assessing a risk, there is safety in numbers. One's faulty perception of a risk could be corrected by another person's clearer perception of an issue. Perception also depends on knowledge and experience. A group will usually have more to contribute than an individual.

Perception and the Limitations of Human Performance

Even when we have achieved perfection in the realm of perception – and this is very unlikely – we still have to put the solution into effect. As human beings, we have limitations in knowledge, strength, physical and mental ability. We have plenty of excuses for getting things wrong.

The major problem is that legislation, the courts, the media and the public at large expect perfection in the realm of health and safety. Representatives of the media will ask: 'Can you guarantee that this will never happen again?' when investigating an industrial accident situation. We can only say something like: 'We have learnt from this mistake and we consider the possibility as now remote'.

Filtering and Selectivity as Factors for Perception

Our senses are continuously receiving information and the brain is processing it. However, we cannot keep our mind on all that is coming to us all the time: we use a filter mechanism.

Information passed to higher levels of management is also filtered. From all the information available, only the vital elements are passed on.

In much the same way, workers are continuously screening out those items that are not of immediate interest. For example, in a noisy workshop, an operator will tend not to hear the background noise. If a person 'speed reads' then they do not see each individual word, rather they quickly scan a page, only seeing words that convey vital information. For instance, if they were looking for some information about 'filtering', they could rapidly scan an article and be stopped by the occurrence of the word 'filter', without reading the whole article.

Filtering and selectivity are vital human activities, since we often tend to do many activities in 'auto-pilot' mode. From a safety point of view, however, the process of filtering and selectivity presents a danger. While concentrating on a particular topic, to the selective exclusion of others, we can easily miss a vital signal which should have warned us of danger. However, we do tend to notice changed situations. Danger signals and warnings are more likely to be noticed if they involve loud bells or klaxons and flashing rather than fixed lights.

STUDY QUESTIONS

3. Outline, with examples, how the human sensory receptors react to danger.

4. Explain how failings in the human sensory and perceptual process may lead to accidents.

(Suggested Answers are at the end.)

Human Failure Classification

IN THIS SECTION...

- Human failure can be classified as errors and violations.

- Errors are actions or decisions that were not intended, and involved a deviation from an accepted standard which led to an undesirable outcome. Errors can be split into three types: slips, lapses and mistakes.

- Violations are a deliberate breaking of safety rules. There are three types of violation: routine, situational and exceptional.

- Rasmussen has proposed three levels of behaviour: skill-based, rule-based and knowledge-based.

- Human error has been identified as a major factor in many serious incidents.

HSG48, Classification of Human Failure

The HSE publication, *Reducing error and influencing behaviour* HSG48, identifies two types of predictable human failure: errors and violations.

Errors

Errors are actions or decisions that were not intended and involved a deviation from an accepted standard which led to an undesirable outcome.

Errors can be further split into several types (based on Rasmussen's skill-, rule- and knowledge-based behaviour theory; see later in this section).

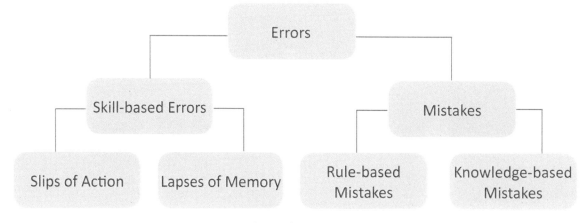

Types of error

TOPIC FOCUS

Skill-Based Errors

(Note: the term 'skill' as used by Rasmussen (and here) is not used in the way people generally understand it.)

These types of error occur in very familiar tasks which require little conscious attention, e.g. an experienced driver driving on a familiar road. Errors can occur when we are distracted or interrupted:

- **Slips** – failures in carrying out the actions of a task. Examples include:
  - Performing an action too soon or too late.
  - Leaving out a step or series of steps from a task.
  - Carrying out an action with too little or too much strength.
  - Performing the action in the wrong direction.
  - Doing the right thing but with regard to the wrong object (or vice versa).
- **Lapses** – forgetting to carry out an action, losing a place in a task or forgetting what we had intended to do. These are often linked to interruptions or distractions. Using a simple checklist can help to reduce the likelihood of lapses occurring.

Possible prevention strategies for skill-based errors include: verification checks, such as checklists; feedback, warning signals if wrong action is selected; design of routines to be distinct from each other; and supervision.

Mistakes

These are where we do the wrong thing believing it to be right. The failure involves our mental processes that control how we plan, assess information, make intentions and judge consequences:

- **Rule-Based Mistakes**

  These occur when our programme is based on remembered rules or procedures. We have a strong tendency to try to use or select familiar rules or solutions. Errors occur if:
  - No routine is known that will solve the new situation, so we don't know what to do.
  - We try to apply the usual remembered rules and familiar procedures because of familiarity with similar problems from previous experience, even when they are not appropriate.
  - The wrong alternative is selected, or there is some error in remembering or performing a routine.

Possible prevention strategies include: simple, clear rule sets; system designed to highlight unusual or infrequent occurrences; clear presentation of information.

- **Knowledge-Based Mistakes**

  These may occur in unfamiliar situations where no tried-and-tested rule exists. They are often related to incomplete information being available or misdiagnosis where, when facing new or unfamiliar situations, we are trying to solve problems from first principles. Errors occur when:
  - Some condition is not correctly considered or thought through, or when the resulting effect was not expected or is ignored.
  - There is insufficient understanding or knowledge of the system.
  - There is insufficient time to properly diagnose a problem.

Possible prevention strategies include: training, supervision, use of checking systems, provision of sufficient time and knowledge.

Violations

Violations are a deliberate deviation from a rule or procedure, e.g. driving too fast or removing a guard from a dangerous piece of machinery, both of which increase the risk of an accident. Health risks are also increased by rule-breaking, e.g. a worker who does not wear ear defenders in a noisy workplace increases their risk of occupational deafness.

TOPIC FOCUS

There are **three types of violation**:

Routine

A routine violation is the normal way of working within the work group and can be due to a number of (sometimes overlapping) factors, including:

- Cutting corners to save time and/or energy – which may be due to:
    - Awkward, uncomfortable or painful working posture.
    - Excessively awkward, tiring or slow controls or equipment.
    - Difficulty in getting in or out of maintenance or operating position (posture).
    - Equipment or software that seems unduly slow to respond.
    - High noise levels that prevent clear communication.
    - Frequent false alarms from instrumentation.
    - Instrumentation perceived to be unreliable.
    - Procedures that are hard to read or out of date.
    - Difficult to use or uncomfortable personal protective equipment. PPE.
    - Unpleasant working environments (dust, fumes, extreme heat/cold, etc.).
    - Inappropriate reward/incentive schemes.
    - Work overload/lack of resources.
- Perception that rules are too restrictive, impractical or unnecessary (particularly where there has been a lack of consultation in the drawing up of the rules).
- Belief that the rules no longer apply.
- Lack of enforcement of the rules (e.g. through lack of supervision/monitoring/management commitment). In certain cases, the violation may even be sanctioned by management 'turning a blind eye' in order to get the job done (related to cutting corners, see above).
- New workers starting a job where routine violations are the norm and not realising that this is the incorrect way of working. This in itself may be due to culture/peer pressure or a lack of training.

(Continued)

TOPIC FOCUS

Situational

Situational violations are where the rules are broken due to pressures from the job such as:

- Time pressure.
- Insufficient staff for the workload.
- The right equipment not being available.
- Extreme weather conditions.

Risk assessments should help to identify the potential for such violations as will good two-way communications.

Exceptional

Exceptional violations rarely happen and only occur when something has gone wrong. To solve a problem, employees believe that a rule has to be broken. It is falsely believed that the benefits outweigh the risks. Means of reducing such violations could include:

- Training for dealing with abnormal situations.
- Risk assessments to take into account such violations.
- Reduction of time pressures on staff to act quickly in new situations.

HSG48 provides a powerful model showing the type of human errors and violations that can be predicted from consideration of organisational, job and individual factors. Such a model can be used both in risk assessments and accident investigations to suggest the control measures required to prevent either an occurrence or a recurrence.

MORE...

You can download *Reducing error and influencing behaviour* HSG48 from the HSE website, at:

www.hse.gov.uk

Cognitive Processing

Decisions have to be made during any working situation; these decisions can be regarded as on-line and off-line processing:

A work process in operation

- **On-line processing** involves those decisions which have to be made as a work process is in operation. Since the human brain can only really deal with a few matters at the same time, operations and the decisions involved tend to be grouped. For example, a machine will be set up to perform a sequence of operations. Once set in motion, it may be difficult to stop the operation until the sequence has been completed. On-the-spot decisions of this type have to be made without too much thought, and so tend to be skill-based. A wrong decision or a missed danger signal can lead to situations where the condition is quickly made worse. Trial and error involves on-line processing.

- **Off-line processing** involves those decisions which can be made after consideration of a number of alternatives. It is often possible to consider, and reject, unsuitable alternatives without the need to try them out first. Often this will involve knowledge and intelligence. Problems occur when we assume that we have correctly interpreted the data available and come up with a solution to the situation. We then fail to search for alternatives, and opt for a wrong course of action. Other errors occur when we attempt to solve a complicated problem mentally, when really it requires a more detailed, written-down, mathematical treatment, or a group decision might be more sensible. Our mental capacity not only depends on knowledge, intelligence and ability, but also on our fatigue levels and our mental state at the time. It is not easy to make correct decisions under situations of pressure, stress or panic.

Knowledge-, Rule- and Skill-Based Behaviour (Rasmussen)

TOPIC FOCUS

Rasmussen's model suggests three levels of behaviour that explain the human error mechanisms:

- **Skill-based behaviour** describes a situation where a person is carrying out a tried-and-tested operation in automatic mode. A competent cyclist can ride a bicycle without any conscious effort or an experienced driver can change gear without thinking of the sequence of events involved. Little or no conscious thought is required; in fact, thinking about the task makes the task less smooth and efficient and increases the chance of error. In this situation, errors occur if there are any problems such as a distraction.

- **Rule-based behaviour** is at the next level – the situation where the operator has available a wide selection of well-tried routines (i.e. rules) from which appropriate ones can be selected to complete the task, i.e. if X happens, then I do Y. An example is obeying *The Highway Code* when driving; if there is a red traffic signal, the rule is to stop. In this situation, errors occur if the wrong rule is applied.

- **Knowledge-based behaviour** is for situations where a person has to cope with unknown situations, where there are no tried rules or skills. The individual, using their experience and perhaps trial and error, tries to find a solution to solve a novel situation. In these circumstances, the chance of error is the greatest.

Contribution of Human Error to Serious Incidents

No study of human error will be complete without some consideration of major incidents. In each case, you need to consider the part played by human error. An investigation usually apportions blame, but although some blame will be attached to those who are directly involved, the majority of the blame is usually placed on those in responsible management positions. Safety practitioners usually carry some responsibility as well.

Kegworth Air Disaster

A Boeing 737 airliner had taken off on a routine flight from London to Belfast. There had previously been a problem with vibration in the right-hand engine, and the maintenance log showed that this had been attended to. The pilot had read the maintenance log before take-off. The air-conditioning on a 737 is driven mainly from the right-hand engine. During the flight, the pilot detected vibration and an excess of smoke and fumes, and he throttled back the right-hand engine. This stopped the smoke and vibration. This seems to have been just a coincidence but seems to have confirmed in the pilot's mind that he had dealt with the problem. The right engine was shut down. There is some suggestion that a warning light showed that there was a fire in the right-hand engine. The pilot obtained permission to land at East Midlands Airport near Nottingham and Derby. It was then found that the wrong engine had stopped: the problem was in the opposite side engine which had suffered a turbine blade detachment. The pilot would have had little difficulty landing the plane with one engine, but now had to attempt to land it on one faulty engine. They crash-landed 900 metres short of the runway on the M1 motorway near the village of Kegworth. There were fatalities and injuries.

The key factors leading to the disaster included:

- The crew did not deal with the initial engine problem in accordance with what training they had.

- Other crew on board observed the flames from the left-hand engine but did not inform the flight crew.

Piper Alpha North Sea Oil-Rig Explosion

On 6 July 1988, there was a disastrous fire on the Piper Alpha oil rig in the North Sea; 167 men were killed and many who survived were injured and traumatised.

The rig was operated by Occidental Petroleum (Caledonian) Limited. Piper Alpha was part of a linked operation involving four rigs. The operation involved gas, compressed gases and crude oil. The various operations on Piper Alpha were in modules which were stacked on top of each other. The helicopter landing pad was on the highest level, and on top of the main accommodation module.

167 men were killed in the Piper Alpha explosion

There were 226 men on the platform; 62 were working the night shift, and the majority of the others were in the accommodation modules.

At 22.00 hours, there was an explosion followed by a fireball that started from the west end of B module. This was quickly followed by a series of smaller explosions. The emergency systems, including fire water systems, failed to operate. Three Mayday calls were sent out, and the personnel assembled on D deck. The radio system and the lighting then failed.

At 22.20 hours, there was a rupture of the gas riser of the Tartan supply (another rig – but the pipeline was connected to Piper Alpha), followed by another major explosion, with ignition of gas and crude oil.

At 22.50 there was a further explosion with a collapse of much of the structure.

There was a mass of photographic evidence, taken from the other rigs and ships in the area, but some problem in fixing the exact time of each. The enquiry was very thorough, but unable to come up with clear conclusions. Gas detection equipment was working, but some water systems had been turned off, and some welding operations were in progress. The report criticised the platform design, and the lack of safety systems. It called for major changes in disaster planning and auditing.

The key factors leading to the disaster included:

- Failure in the permit-to-work system.

- Design failure in that the rig containment wall was fire-resistant but not blast-resistant.

- Other rigs did not shut down and continued to feed into Piper Alpha, fuelling the fire.

- Inadequate emergency procedure for rig evacuation.

Herald of Free Enterprise

The *Herald of Free Enterprise* sailed from Zeebrugge harbour for Dover with both inner and outer bow doors open. Water flooded in, causing the ferry to capsize. The Assistant Bosun was responsible for closing the doors but had fallen asleep. The Captain assumed that the doors were closed unless told otherwise. There was pressure on ferries to sail as quickly as possible.

The key factors leading to the disaster included:

- Design failings, in that roll-on, roll-off ferries were inherently unsafe and top-heavy.

- Reduction in the complement of officers, with long working schedules.

- No automatic monitoring of critical areas such as the bow doors.

- Poor emergency procedures, particularly provision of lifejackets.

Ladbroke Grove (Paddington Rail Disaster)

On 5 October 1999, a local passenger train passed a red signal and continued into the path of a high-speed train. The ensuing collision and subsequent fire resulted in the deaths of 31 people, with many injured.

A joint investigation by the Health and Safety Executive (HSE) and British Railway Police identified a number of significant problems associated with the signalling system.

Among these was the positioning of a particular signal that was exceptionally difficult to read in comparison with other signals. It was also suggested that the driver's perception could have been affected by the sun reflecting on the signal lenses.

Additionally, there was some debate about whether the Automatic Warning System (AWS) could have given misleading warnings which led to them being disregarded.

One of the main conclusions was that the misinterpretation of the information presented by the signal was a significant factor. The competence of the driver was not questioned as they had been fully trained, although they were relatively inexperienced.

Three Mile Island (USA)

Three Mile Island nuclear power plant is near Middletown, in Pennsylvania, USA. In March 1979, the reactor core went into meltdown, but there were only small releases of radioactivity and no deaths/injuries. The reactor was a Pressurised Water Reactor (PWR), which is a very common design throughout the world (other than in the former Soviet Union). In this design, the primary coolant water is kept pressurised so that it does not boil. The primary coolant water passes the heat onto a secondary water system (via a heat exchanger), which is allowed to boil – the steam driving a turbine to generate electricity.

To summarise, the incident started with a failure of the secondary circuit – which prevented heat removal. This caused the reactor to shut down and the pressure in the primary circuit to increase. This in turn triggered the opening of a pressure relief valve which, unfortunately, stuck open instead of closing again when the pressure had reduced. As it happened, the signals on the operating consoles indicated that the valve had shut (the lamp was triggered by the circuit signal to the valve rather than the actual valve position). The continued escape of the coolant through the valve allowed the core to overheat.

Coupled with this was the confusing instrumentation available to the operators. There was no coolant level indicator – instead it was inferred from levels elsewhere in the system (but these levels had actually been raised by bubbles of steam). Alarms began to sound but, at that stage, the nature of the unfolding incident was not recognised as a 'loss of coolant' incident. Immediate actions included reducing the coolant flow in the core (their training had emphasised the danger of too much coolant and, of course, they believed, from the instrumentation, that the pressure relief valve was shut); this made things worse. If they had done nothing, the plant would have cooled down on its own. Instead, the core continued to overheat and the fuel began to melt.

Factors that led to the accident:

- Operators were under considerable stress – many alarms were going. They had incorrectly diagnosed what they thought was the problem and stuck to a course of action, despite apparently overwhelming evidence to the contrary.

- Operator training was inadequate. Operators of complex plant cannot just be given a series of instructions to follow. Things are bound to go wrong outside of this. They also need to understand the principles of the process, and be trained in diagnosing problems (both foreseen and unforeseen) and in the use of diagnostic aids.

- The crucial indicator (of the status of the pressure relief valve) was wrong. This did not look at the status of the relief valve directly – it should have done.

STUDY QUESTIONS

5. According to Rasmussen's model, what are the three levels of behaviour?

6. How did human error contribute to the Piper Alpha disaster?

(Suggested Answers are at the end.)

Improving Individual Human Reliability in the Workplace

IN THIS SECTION...

Human reliability may be improved by:

- Incentive and reward schemes.
- Increasing job satisfaction.
- Appraisal schemes.
- Selecting appropriate individuals for the job:
    - Matching skills and aptitudes.
    - Training and competence assessment.
    - Ensuring fitness for work.
    - Health surveillance.
    - Support for ill health, including stress-related illness.

Motivation and Reinforcement

Workplace Incentive and Reward Schemes

Workplace incentive or reward schemes can be a good way of motivating employees to focus on the job and conform to the organisational goals. The incentive encourages employees to work harder in order to receive a payment or benefit. For example, achieving a set target or exceeding that target may mean that individuals receive a financial bonus or a prize. The scheme may operate on an individual basis or as a team effort, in which case the incentive would be for the team to achieve the target. The incentive scheme may be set up so that a winning team or individual is identified every month, for example, and the winner is given a prize. This type of incentive motivates individuals to work harder but also motivates teamwork.

Some pay schemes work by paying a very low actual salary but having bonus payments which are paid when targets are met, e.g. sales jobs. The motivation to sell more is clearly through the financial gain.

Piecemeal work, i.e. where workers are paid per work unit completed (e.g. for each sheep sheared) encourages individuals to work quickly so that they can earn more money. This may have implications with respect to safety as the workers are not encouraged to necessarily consider safety as their first priority.

> **DEFINITION**
>
> **INCENTIVE**
>
> An incentive is really an inducement that provides a motive for someone to do something, usually in the form of some sort of reward for achieving a particular goal or milestone.
>
> (Note that the words 'incentive' and 'reward' in this context are routinely used interchangeably.)

Incentive schemes aimed at improving safety are more difficult as they may need to monitor the results over a reasonable time period, e.g. one year. Incentive schemes can often lose their momentum and their effectiveness over time. For this reason, it is important to either keep the time periods short or continue to keep the momentum high.

Incentive schemes for safety may relate to obtaining improved 'scores' during routine audits or inspections. This type of incentive would be aimed at ensuring all members of the workforce made their work area as safe as possible and that work was carried out in a safe manner.

Avoid incentives aimed at reducing accidents specifically, i.e. where measurement would be a decrease in accident rates, as this may result in under-reporting of accidents in order to obtain the incentive.

Job Satisfaction

For some people, job satisfaction is all that they require to be motivated. Job satisfaction is also very individual to each person:

- Some people are satisfied with a good working environment and regular rest breaks.

- Other people require challenging, stimulating work where they receive positive feedback.

One motivation theorist, Herzberg, identified particular motivating factors which, when present, increase satisfaction from work and provide motivation toward superior effort and performance. These include recognition, responsibility, achievement, advancement and the work itself, and are distinct from other factors that increase dissatisfaction when absent, but when present do not result in positive satisfaction and motivation. Herzberg termed these 'hygiene' factors. They include type of supervision, salary/wages, working conditions, company policies, rules, etc.

Workers on a production line

Appraisal Schemes

Appraisal schemes usually involve the employee filling in a self-appraisal form which is discussed at an interview with their manager. A report is produced at the end of the interview with a copy being provided to the employee and to a senior manager, and a copy placed on the employee's personal file.

The self-appraisal form may request information about what the employee feels they have accomplished in the past year and their high and low points. It may also ask what areas the employee is dissatisfied with and what improvements they would like to see. The form may also ask about the employee's aspirations for the coming year. In this way, the employee is given an opportunity to identify what areas of their job they are satisfied with and what areas they are dissatisfied with. They may also come up with ideas to improve their job or to improve themselves, e.g. additional training. This scheme gives the manager an opportunity to discuss with the employee their thoughts on the employee's progress, and give praise and encouragement where required.

> **DEFINITION**
>
> **APPRAISAL SCHEME**
>
> A formal means of placing value on achievement or effort and is generally carried out on an annual basis. The results may be used to determine the level of a pay rise or a promotion.

Some appraisal schemes give the employee the opportunity to comment on their manager. This needs to be anonymous if there is a chance of reprisal.

Appraisal schemes are an excellent way of finding out what problems exist within a workplace and, therefore, give the opportunity for improvement. They also provide a measure of the safety culture within an organisation. More importantly, they allow the employee to comment on their own progress and to voice their opinions. Employees in appraisal schemes will often feel more motivated than those not in such a scheme, particularly where hard work and improvements are rewarded.

Selection of Individuals

Matching Skills and Aptitudes

An employer may wish to select only those workers who will conform to their safety standards – either existing workers for new or different tasks, or prospective employees. This selection process is often by interview (at least in part) and sometimes involves aptitude tests.

Be aware that selection in this way may not lead to large improvements in reliability because people may behave differently once they have the job.

Some of the best selection techniques involve **competency-based interviewing** which identifies the skills, talents and abilities required by the job and may assess:

• Effectiveness of communication (verbal and written).

• Problem-solving ability.

• Ability to use technology.

• Whether they provide input or views on safety issues.

• Whether they follow safety instructions.

• Teamwork, etc.

Training and Competence Assessment

• **On-the-Job Training**

 On-the-job training provides trainees with experience which is a combination of work-based knowledge and skills development. As the trainee gains experience, the range and complexity of tasks that they can undertake without detailed guidance increases. The process of learning can be improved by:

 – Demonstration.

 – Coaching (carrying out tasks with guidance).

 – Projects.

 The instructor assesses the competence of the trainee as their skill level increases.

 This training is effective, providing the trainee is shown the correct way of carrying out the task; bad habits can develop from the start if the trainee is placed with someone who does not follow the correct procedures.

• **Off-the-Job Training**

 Off-the-job training is carried out away from the work environment in a number of ways:

 – **Lectures** – one-way communication in which all the talking is done by the lecturer. It is a good way of teaching a large number of students simultaneously. The limitations are:

  – There is a very low rate of retention. Adequate back-up notes are essential.

  – Students may not understand the presentation and be unable to seek clarification.

 – **Seminars** – where discussion is encouraged and students can learn from the instructor and from each other. The number of students who can usefully take part at one seminar is a limiting factor.

 – **Programmed instruction** – provided through a combination of distance learning or open learning packs, computer or audio-visual programmes with no direct involvement of an instructor. However, many distance learning packages do have access to tutors for advice or assistance through e-mail or telephone contact.

Fitness for Work

Some jobs, often called 'safety critical', involve activities that require a person's full, unimpaired control of their physical and mental capabilities. For example, a tower crane operator will need:

Good eyesight is essential for a crane operator

- To be able to climb safely up the mast to the cab.

- Good eyesight.

- To not suffer from any condition that might make them prone to lose consciousness.

In such circumstances, the employer would require a full medical assessment of the prospective employee by an occupational health practitioner. Although the results of the assessment are confidential to the worker and the practitioner, the employer can expect to be given a general report specifying whether the prospective employee is:

- Fit for work.

- Fit for work with restrictions.

- Temporarily unable to meet the fitness standard.

- Unable to meet the fitness standard for work to carry out specific jobs.

Health Surveillance

Health surveillance involves implementing systematic, regular and appropriate procedures to detect early signs of work-related ill health among employees exposed to certain health risks and then taking appropriate action. Examples include hearing tests, lung function tests and blood tests for substances such as lead.

Regulation 6 of the **MHSWR** requires that: *"Every employer shall ensure that his employees are provided with such health surveillance as is appropriate having regard to the risks to their health and safety which are identified by the assessment."*

For most employees, employment health surveillance is not necessary. However, guidance from the HSE suggests that health surveillance will be required:

- when there is a specific legal requirement; and

- where:

  - there is an identifiable disease or adverse health condition related to the work concerned; and

  - valid techniques are available to detect indications of the disease or condition; and

  - there is a reasonable likelihood that the disease or condition may occur under the particular conditions of work; and

  - surveillance is likely to further the protection of the health and safety of the employees to be covered.

MORE...

Specific information on health surveillance is available at:

www.hse.gov.uk/health-surveillance/what/index.htm

Support for Ill Health and Mental Health Problems

Providing support for workers suffering from ill health (including work-related mental ill health) will be beneficial for both the worker and the employer. A good occupational health service will advise on a rehabilitation programme so that the worker can re-adjust to the work environment and not jeopardise their recovery. For example, an individual who has suffered from work-related mental ill health, such as symptoms of anxiety or depression (whether or not it has been caused or made worse by the work) is likely to benefit from working part-time before returning to full duties.

STUDY QUESTION

7. Identify four different methods by which employees can be motivated.

(Suggested Answer is at the end.)

Organisational Factors

IN THIS SECTION...

- Human error is increased by inadequacies in:
    - Policy.
    - Setting of standards.
    - Planning.
    - Information.
    - Responsibilities.
    - Monitoring.
- In any organisation there are both formal and informal structures.
- Communication mechanisms within an organisation vary in their complexity, reliability and formality.

Effect of Weaknesses in the Health and Safety Management System on the Probability of Human Failure

Inadequacies in Policy

Organisations must appreciate that they need to consider human factors as a distinct element which must be recognised, assessed and managed effectively in order to control risks. For this reason, human factors must be considered and included in the company health and safety policy; if they are not, then it is more likely that these important factors that can affect the way in which individuals work will be overlooked and human failure is more likely to occur.

For example, some of the common organisational causes of human failure include:

Human factors must be considered in the health and safety policy

- Inefficient co-ordination of responsibilities.
- Poor management of health and safety.

Both of these stem back to inadequacies in policy. If the health and safety policy defines the responsibilities correctly and the ways in which health and safety are to be managed, then failure is less likely to occur due to these causes.

Setting of Standards

The setting of standards and the use of benchmarking is a feature of any health and safety management system. Recognising human error is essential in such areas, as identifying foreseeable misuse is a necessary element of a suitable and sufficient risk assessment.

Information

The availability of information within an organisation or system is vital and the information should be:

*   accurate,

*   timely (e.g. it is no good being informed of a new procedure three weeks after the implementation date), and

*   relevant.

Too much information can be overwhelming

Too much information can be overwhelming and will mean that the important bits may get overlooked.

Providing the right information, at the right time, and to the right people, is not easy but it goes a long way to ensuring a good working system and one where the employees feel involved and appreciated. One example is ensuring written instructions (including warning signs) are clearly understood by everyone, including those with a poor understanding of English. Anyone who has worked for a company where information was not provided adequately knows the confusion and mistrust this can cause.

The information required may range from the structure of the organisation and the responsibilities within it, to the operating instructions for a piece of equipment.

Planning

The proper planning of a system ensures that it works effectively, and so all aspects of it must be taken into account. This includes the inputs, the outputs, the work in the middle (production/processing), as well as the effect of the environment. All these areas need to be looked at to see how they affect the system or how the system affects them. Different scenarios should be considered so that the system can operate in changing circumstances.

For efficient working, system planning must take account of relationships between processes, i.e. the organisation and communications, and the ability to adapt to change. This might include:

*   Proper work planning, including the task steps as well as relationships with other tasks – to remove unnecessary work pressure.

*   Properly integrated procedures and safe systems of work.

*   Proper co-ordination.

*   Communications – two-way to allow feedback for improvements and clarification.

Responsibilities

To implement an effective system, everyone involved must understand their role and how it integrates into the system. Each person must also appreciate the effect on the system as a whole if they don't play their part. Unless responsibilities are clearly defined and understood, then there will be an increased risk of tasks not being fulfilled, e.g. maintenance. This will have a consequential effect on safety and health.

Monitoring

Feedback and monitoring of a new system is vital to ensure that the system works and that, where necessary, improvements are made. Human error is significantly reduced by providing proper, timely feedback to the individual or group.

Influence of Formal and Informal Groups

Formal Groups

You will remember that we looked at formal and informal organisations in an earlier element.

Formal organisations:

- Are established to achieve set **goals**, **aims** and **objectives**.

- Have clearly defined **rules**, **structures** and **channels of communication**.

- Are often divided into **productive** and **non-productive**, productive organisations being involved in the production of goods and services.

To be successful, an organisation has to have clearly defined objectives and be positive in aiming to achieve them in the most efficient manner. Where this positive direction is lacking, an organisation is likely to fail.

Nearly all organisations are **hierarchical** in structure, i.e. they have different levels of authority and responsibility within their structure.

The simplest way of depicting such a functional hierarchy is with a **line diagram** (or organisation chart) similar to the one that follows.

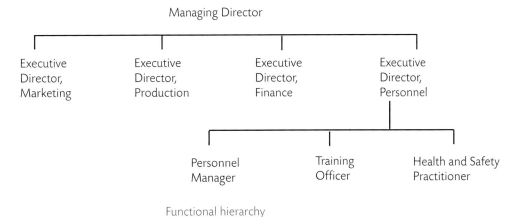

Functional hierarchy

Organisations also make use of matrix charts to depict organisational structure. In the figure below, staff functions are shown across the top and line functions down the side. Interaction takes place where the functions cross.

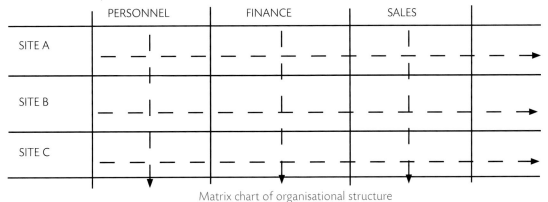

Matrix chart of organisational structure

Concentric circle charts (see the following figure) show the management functions to be the hub of the organisation around which all other decisions and functions revolve.

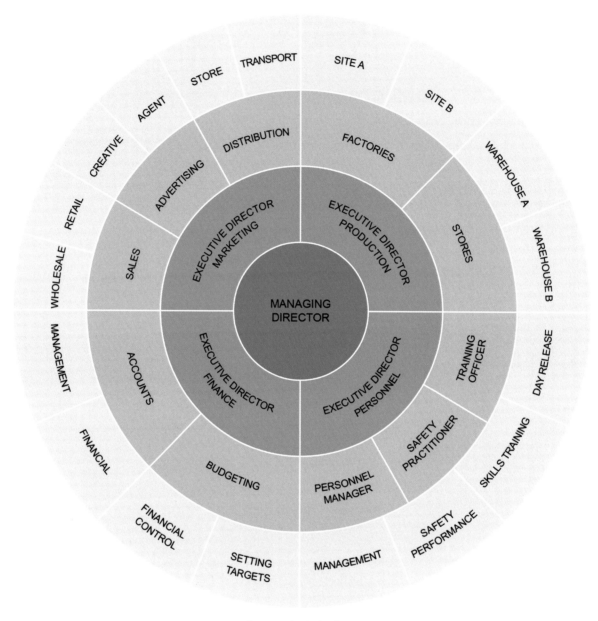

Concentric circle chart

Informal Groups

The organisation chart shows the formal organisation of the company and indicates the direction of communications. There will also be formal working groups and committees. In a large organisation, this can be cumbersome and some decision-making processes use informal routes. The health and safety practitioner needs to be aware of these informal methods.

Although a formal structure would suggest that they might communicate with the works foreperson by reporting to the human resources manager, who contacts the managing director, who then passes the information or instruction to the foreperson via the production manager and supervisor, in practice the safety practitioner goes direct to the foreperson and, if necessary, reports this using the formal channels.

Within any organisation there is a 'grapevine'. This is usually very effective in passing on gossip and information. Since the source is difficult to trace, the information might not be totally reliable. So, superimposed on the formal organisational structure is an informal structure of communication links and functional working groups. These cross all the barriers of management status and can be based on:

- Family relationships.

- Out-of-work activities, such as the church, golf club, or local pub.

- Valuable experience or expertise.

Organisational Communication Mechanisms and Their Impact on Human Failure Probability

Every organisation depends upon an intricate communication network; the bigger the organisation, the more elaborate the system. The precise form of the network will vary from company to company. The following networks and the direction of communication they deal with are the most common, although, in practice, the communication system in a particular company will inevitably be far more complex. Communication systems vary in their complexity, reliability and formality.

Modes of Communication

Communication can be either one-way or two-way.

In **one-way communication**:

- Sender identifies the message.

- Sender transmits the message.

- Receiver receives the message.

- Receiver interprets the message.

Although this is quick and gives the perception of efficiency and control, there is no opportunity for feedback and the assumption is that the receiver has paid adequate attention.

Examples include: a tannoy message in a factory, a safety poster, following written or e-mail instructions.

In **two-way communication**, there is the opportunity for the receiver to transmit information or questions back to the original sender and for the sender to respond such that a conversation takes place. Although more complex and time-consuming, two-way communication is likely to be more effective and reliable by placing the onus on both parties rather than one. Achieving a mutual understanding between the two parties ensures that the correct message is received and understood and contributes to an improved safety culture.

Examples include: a one-to-one meeting, a toolbox talk with the opportunity for questions, etc., or a telephone call.

Shift Handover Communication

Shift-working and shift handovers are characteristic of many organisations, not only in the process industries but also in healthcare. During shift handover, relevant information has to be communicated to maintain the continuity of the activities; if this fails, there is the risk of serious consequences. A key factor in the Piper Alpha disaster in 1988 (discussed earlier) was the failure in the permit-to-work system such that the oncoming shift members were unaware of the removal of a safety valve. This failure led to actions that initiated the disaster.

In 2004, the British Medical Association published guidance on good practice handovers in healthcare and detailed five questions:

- Who should be involved? – All key personnel at all grades.

- When should handover take place? – At fixed times, of sufficient length and arranged to allow both the off-going and on-going shifts to attend within their working hours.

- Where should it take place? – Close to the most used work sites so that there is room for both sets of staff to attend.

- How should handover happen? – A specific formal format should be devised and consistently followed.

- What should be handed over? – This might include written notes as well as electronic information.

Organisational Communication Routes

Vertical Communication

The amount of communication downwards tends to exceed that going upwards:

- **Downwards**

  Communication will usually be made along the lines of authority, from Managing Director down to Section Leader and on to the clerk and shop-floor worker (see following figure).

Communication routes

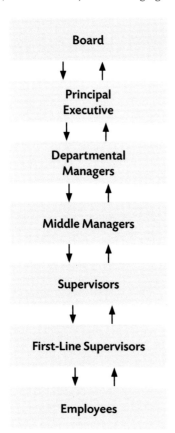

Vertical communication – downwards and upwards

At each managerial level, there must be responsibility for passing on information. Each superior level must be responsible for ensuring full, accurate knowledge and understanding at the next subordinate level. The importance and use of communication must, therefore, be included in any management training programme.

The passing 'downwards' of some directive, communication or instruction, implies temporary 'storage' of that information in the mind, or the 'in-tray', of all intermediate handlers. Careful consideration must be given, therefore, to the most appropriate type of information storage and display system.

Some senior staff believe that the only effective way to pass information is by word of mouth – their mouth! They think they are the only really effective communicators in the organisation, but this can mean that they find themselves with no time to make decisions because all their time is taken up ensuring that the decisions they have made have been passed on to 'all concerned'.

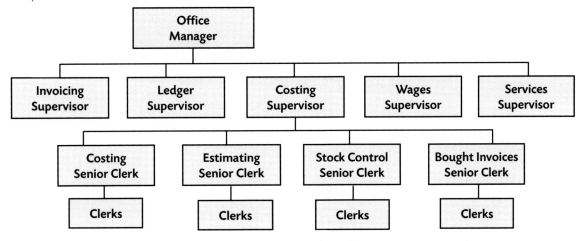

Vertical communication within a department

- **Upwards**

  Communication upwards is equally important in any organisation – ideas, suggestions for improvements, and opinions on existing systems, communications and techniques are all important for management to consider and use.

  Office and shop-floor workers are in direct contact with the actual work carried out and can often see ways to improve processes and production. The regular flow of such ideas has been of considerable value to organisations in reducing costs, cutting production times, introducing improved layouts and in creating an atmosphere of co-operation and goodwill between employees and management.

  Research has shown that it is in the upward flow of information that the greatest shortcomings exist, especially in recent years, with the use of management information systems and the selection and processing of the 'vital' information managers need to have.

  Whereas downwards communications are usually 'directives', i.e. they initiate action by subordinates, upward ones are usually 'non-directive', i.e. they report results or give information, but are not necessarily intended to prompt action.

  Although the amount of downwards communication is usually greater than that going up, managers should encourage an increase in the flow upwards, although much depends on the time the manager has available to deal with the upwards communications.

Horizontal Communication

Information is also channelled horizontally, both within a department and between departments.

We give information to and receive it from colleagues in our own department and we have contacts with our opposite numbers in other departments. These communications are of the greatest value in administration, particularly in affecting co-ordination (see the following figure).

Remember that information flow is subject to variation in speed and quantity; activity will vary according to the time of day, the day of the week, and the month or quarter.

The characteristic of **feedback** is vital in effective communications. It should inform the sender of information that their message has been understood and acted upon, hopefully in the expected manner, bringing about the planned objective.

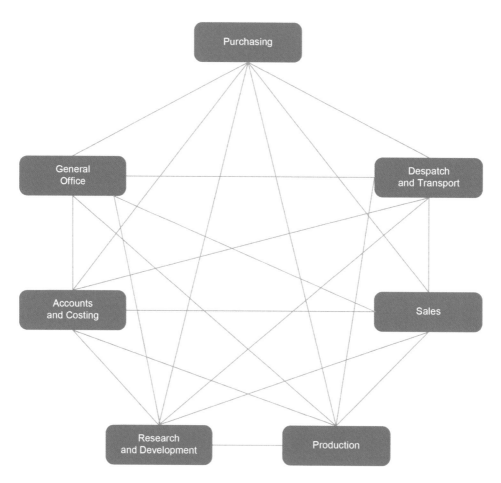

Internal communications – horizontal

Inward and Outward Communication

- Inward

  Here, we see the effect of all the personal face-to-face calls on people at all levels in the organisation: the incoming telephone calls and e-mails from people of all kinds making contact with various members of staff, and postal correspondence arriving daily in the post room.

- **Outward**

The amount of communication outwards from any organisation is sometimes grossly miscalculated. Outgoing communications are both formal and informal, both explicit and implicit.

In this same category, we can include the behaviour of responsible members of staff when they are off company premises; their behaviour and expressed attitudes may be seen as reflecting those of their organisation.

Outward communications also include the various kinds of advertising and promotional devices the organisation uses.

STUDY QUESTION

8. Outline the differences between formal and informal groups within an organisation.

(Suggested Answer is at the end.)

Job Factors

IN THIS SECTION...

- Job factors have a major influence on human error, including:
    - Task complexity.
    - Patterns of employment, e.g. short-term contracts.
    - Payment systems, e.g. piece work.
    - Shift work.
- Task analysis breaks down the job into individual steps, which can be further analysed.
- Ergonomics is the study of adapting the job to the individual.
- Poorly designed workstations can increase human error as well as cause ill health.
- Physical stressors (e.g. extremes of temperature), fatigue and stress, all have an adverse effect on human reliability.

Effect of Job Factors on the Probability of Human Error

Job factors include such things as:

- The equipment, e.g. design and maintenance of displays, controls, etc.
- The task itself, e.g. complexity.
- Workload.
- Procedures or instructions – clarity, completeness.
- Disturbances and interruptions.
- Working conditions – noise, temperature.

Task Complexity

The complexity of the task can have a significant effect on the propensity for human error. Tasks that involve complex calculations, decisions or diagnoses will present more opportunity for such error. Such tasks should be broken down into simpler units to give greater clarity.

Patterns of Employment

These days, it is accepted practice for workers to be on short-term contracts and this clearly has an effect on individuals seeking job security. Some people in this position may suffer from stress due to the lack of job security, particularly if they have always had a 'permanent' job in the past. Other people acknowledge that there are now few 'jobs for life', and take short-term contracts in their stride.

Human error may be more likely with complex tasks

Short-term contracts often mean that the employer can choose to retain the best workers. Where there is a good safety culture in place, this will often mean workers who perform well and safely. Where the safety culture is poor, this may mean workers who work the quickest will be kept on. So, the type of organisation will determine how the individual will work in order to ensure that their contract is renewed. On the other hand, where short-term contracts are in place, there may be little loyalty from the workforce and so turnover of staff (particularly good performers) may be high.

'Permanent' contracts may lead to complacency in the workforce, in which case the employer needs to ensure that individuals achieve their potential and work toward the company goals. There are many ways of encouraging improved performance, e.g. reward and incentive schemes.

The way the work is organised between people can also have a major effect on performance. Where people work in small teams with some variety to their tasks then this can build comradeship and a good working environment. However, where people work alone, work can become a lonely place and the tasks can become monotonous.

Payment Systems

The way in which people are paid can have an effect on the way they work. For example, piecemeal workers are paid by performance; abattoir workers are often paid per animal slaughtered, so for them, speed is of the essence because, the faster they work, the more they get paid. While safety may not be the top priority, they understand that their own safety is paramount because if they injure themselves, they won't be able to work and then they won't get paid. So, by default, personal preservation may lead them to work more safely.

This is really the same for all self-employed people. On the other hand, employed people who still get paid if they are absent from work may not think about their own safety in these terms, and so may or may not work more safely.

Shift Work

Shift work has a great effect on an individual's performance. In addition to fatigue and stress, individuals may find that their social lives and family life are affected. The effects of this will rather depend on the individual and their circumstances, as well as the shift pattern itself. If, however, an individual is unhappy at home, then this will often spill over into their work life and performance may be affected.

Shift workers (especially night workers) may experience negative effects on their health:

- Gastrointestinal problems are more likely to occur due to eating snack meals during work hours.

- Respiratory problems, such as asthma, tend to be worse at night, as do allergic reactions.

- Lung function also declines at night, especially for those people with chronic respiratory problems. Clearly, where people's health is affected, performance may also be affected.

Shift work interferes with the body's natural circadian rhythm. Even when working nights, the body still reduces it's temperature in the early hours of the morning, reduces blood pressure and stops digestion, which leads to an individual feeling sleepy and less alert.

Shift workers need adequate rest between shifts as well as regular rest days to 'recharge their batteries'. The shift pattern itself may also affect individuals. Shift patterns that alter once a week are likely to be more difficult to adjust to rather than those that change more rapidly or more slowly.

Application of Task Analysis

Task analysis is a process that identifies and examines tasks performed by humans as they interact with systems. It is a means of breaking down a task into each individual step and is a technique that looks at an activity in detail. The activity in question may be one where a number of people have injured themselves. By breaking the task down into each step, then the cause of the injury may become apparent and a better way of completing the task may be identified. Each step can be examined in detail to try to identify where human error might occur. Could slips occur by performing an action too soon or by leaving out a step from a task? Is it possible to make a mistake by selecting a wrong alternative? Once the possibility of human error has been identified, the task can be modified to reduce its likelihood.

Task analysis

For example, consider the school cooks who produce hundreds of hot meals per day. A number of incidents have occurred, which on the surface seem unrelated, e.g. back strain, burns, slips and trips. Looking more closely, however, it seems that all the accidents have occurred while removing or putting items in the oven. After breaking down each step, it becomes apparent that the oven door does not always stay fully open so the cook has to balance the trays of hot food while trying to keep the door open. This sometimes means holding the food in an awkward manner (leading to back strain) or spilling the food (leading to slips and trips) or being burnt by the oven door. In this simple example, it may be possible just to fix the door so it stays open, or to change the procedures, or it may mean someone has to hold the door open while another person removes/replaces the food.

By breaking down a task, you can see exactly what happens without making assumptions about some of the steps.

Role of Ergonomics in Job Design

Ergonomics is concerned with 'fitting the job to the worker' rather than expecting the individual worker to adapt to the job.

Influence of Process and Equipment Design on Human Reliability

Human beings are unreliable – how unreliable depends on the individual and the work environment. Consider the effect that being in a very hot environment has on your work performance; or, when you have had a large lunch, how your output is affected by a feeling of sleepiness. However, much can be done to minimise such effects by improving the environment and making the task such that errors are minimised. This is achieved by careful design of any controls.

Worker and machine are each better at some things than the other. Ideally, you want to use the strengths of both to minimise possible weaknesses; together they represent the 'system' for meeting the requirements. This can be illustrated diagrammatically as in the following figure.

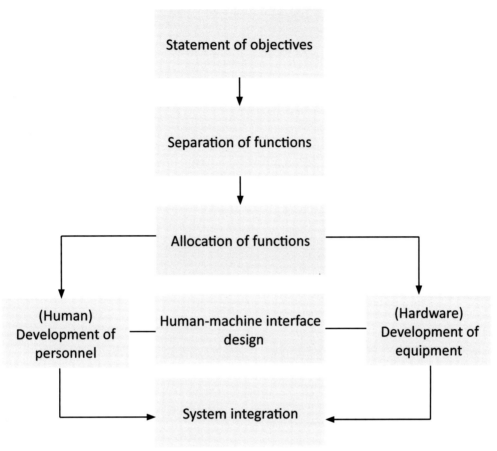

Statement of objectives

↓

Separation of functions

↓

Allocation of functions

(Human) Development of personnel — Human-machine interface design — (Hardware) Development of equipment

System integration

System design process

Ultimately, every piece of plant can be represented by what is often called the HMI, or 'human-machine interface'. Ergonomics is the study and design of this interface, such that the operator can perform their duties efficiently, in comfort and with minimum error.

Grouping of displays and dials next to their controls, and consistency in these displays – for example, all moving the same way for increase – are important in allowing the operator to form a mental picture of what is actually happening in the plant under that person's control:

- Displays should be arranged so they can be scanned with minimum effort.

- Display dials should be the appropriate type for the reading (digital v. analogue).

- Dials should have identified areas for normal and abnormal readings to make it easier to see if something is starting to wander, e.g. a fuel gauge in a car.

- Bulbs and other indicators should be shielded from glare so that their status cannot be confused.

Consistency is important in the action of control devices; we all expect to turn something on and increase some variable, by turning a knob clockwise. Relative positioning of control devices and displays is important. People expect to see a reaction to an action, even if it is only a light that indicates the action is being acted upon (the lift call-button effect) and, of course, they should be within easy reach of the operator. Controls should be organised and laid out so as to logically follow the process. They should be clearly marked or labelled. The number of controls should generally be kept to a minimum.

Displays and dials should be consistent

Emergency arrangements should be distinctive so that emergency stop controls can be easily located, and any audible warnings should take account of expected background noise levels.

Possibly the most important aspect is the worker's immediate working space and environment. Reliable work cannot reasonably be expected from an operator who has a headache or a sore back within an hour of starting work. Factors such as noise, dust, smell, vibration, temperature (and temperature changes), lighting levels (and glare), and humidity all contribute to a worker's ability to concentrate. Psychological factors, such as the degree of concentration necessary, and the ability to mentally rest and 'coast' for a short period, are also important. Also, remember the importance of providing chairs, if appropriate, to avoid fatigue from prolonged standing.

The layouts of controls, displays and seating for convenience of operation are often overlooked.

The appropriate grouping and display of controls is vital

To ensure that the strengths of worker **and** machine are utilised, a Fitts List (named after Paul Fitts, who developed the technique) is produced for the system.

Example of Fitts List

Activity	Machine	Human
Speed	Much superior	Lag > 1 second
Power	Consistent, large, constant, standard and precise forces	2 hp for 10 seconds 0.6 hp for minutes 0.2 hp over day
Consistency	High	Not reliable, must be monitored
Reasoning	Good deductive	Good intuitive
Input sensitivity	Some outside human range (e.g. X-rays)	Wide range, good pattern identification

hp = horse power

Note that this is given as an example for a system that requires the given ability. Each system will require a Fitts List developed to suit its specific requirements for actions to be performed, although, with practice, such a list does not take long to produce.

The Employee and the Workstation as a System

In ergonomics, the worker, the machine and the working environment may be considered as the elements which together comprise a system.

When considering the ergonomic 'fit' of the workplace to the worker, there are a number of factors to take into account.

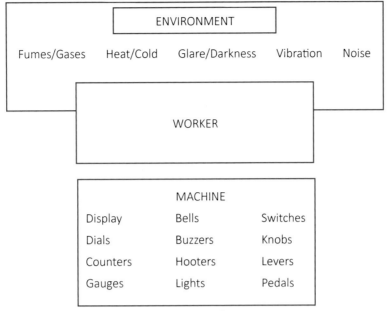

Ergonomic 'fit'

Elementary Physiology and Anthropometry

The skills of an ergonomist include:

- **Anthropometry**

  This is a study of human measurements, such as shape, size, and range of joint movements. A machine must be designed for the person. Since no two people are the same, a design is required that will suit, or can be adapted for, a wide range of sizes of individuals. It is typical to design for a range of two standard deviations from the mean. (Remember that we looked at the statistical terms 'standard deviation' and 'mean' in an earlier element.) This includes all but the extreme 10% at each end of the measurement scale. Group characteristics must also be considered, e.g. the average height of people varies between different populations.

- **Physiology**

  This is a study of the calorific requirements of work (how much energy is needed) and body functions, the reception of stimuli, processing and response. The operator and machine must be complementary. A person must not be expected to do more than the human body is capable of. Some things are best done by a person; other things by a machine.

  Physiology includes a study of the operation of machines. A person can operate two foot controls when sitting, but only one when standing. An investigation by Cranfield Institute of Technology determined the ideal dimensions of the average operator of a horizontal lathe – 'Cranfield worker' would need to be 1.35m tall and have a 2.44m arm span.

The ergonomist should make a contribution at the design stage to try to prevent problems occurring later.

Degradation of Human Performance Resulting from Poorly Designed Workstations

The British Standard 3044:1990, *Guide to ergonomics principles in the design and selection of office furniture*, is one example of the help available to designers. The HSE leaflet INDG90, *Ergonomics and human factors at work*, aims to help employers and managers to understand ergonomics and human factors in the workplace by giving examples of ergonomics problems and simple, effective advice about how to solve them.

Workstations are usually designed for the 'average' person. If a doorway was designed just for the average person, then some of the population would have problems getting through.

Workstations need to be capable of adjustment. Unsuitable workbench height causes the operator to develop musculoskeletal problems:

- If the workbench is too high, the operator has to adopt an unnatural posture, with the elbows away from the body and the shoulders raised. This causes discomfort in the shoulders and neck.

- If the work surface is too low, the operator will have to lean forward. This causes neck and lower back problems.

- Repetitive movements, particularly those requiring the operator to exert force or use an unnatural action, can lead to upper limb disorders. One problem is tenosynovitis or inflammation of the tendons of the hand and wrist. This is a common problem with keyboard operators.

Ergonomically Designed Control Systems

- **Production Process Control Panels**

  The operator of a production process control panel must be able to operate the panel from a safe place. For some production processes, this may be from an adjacent area or, for more dangerous operations, the panel will be located at a safe distance or even within an enclosed area away from the production area. Noise, dust and fumes must all be considered.

  The operator must be able to reach all the dials, switches, etc. easily. Emergency controls must be clearly identifiable and easy to operate. The operator must also have a view of the production area so that they can see what is happening and react, as necessary.

> **MORE...**
>
> The Ergonomics Society, Elms Court, Elms Grove, Loughborough, Leicestershire, LE11 1RG, are able to supply further information on this subject.

- **Crane Cab Controls**

  A crane driver has to be in absolute control of the load that is being moved because the slightest slip of the controls may result in damage to buildings, materials or people. For this reason, it is vital that the controls in the cab are within easy reach and move in straight lines to permit ease and delicacy of control. The driver must be provided with an adjustable seat (to fit accurately 90% of all possible sizes) so that they have a full view of the working area.

  The driver must also be protected from the ingress of dust, fumes and heat from the external environment. The provision of filtered and refrigerated air, where necessary, ensures cool and comfortable working conditions.

A crane driver must have a full view of the working area

- **Aircraft Cockpit**

  It is vital that a pilot can interface easily with all the controls in the cockpit. The controls/displays must be fitted around the cockpit in a logical way so that the pilot can easily reach and see the more important controls/displays, e.g. speed and altitude dials, while they may need to move to reach the less important ones. It is important that safety-critical switches cannot be inadvertently operated. These should be designed so that there has to be a positive action by the operator in order to initiate them.

  Emergency controls must be clearly identifiable, easy to use and situated in a suitable location. The emergency controls must be accessed quickly to prevent unnecessary delay in stopping the activity that they control.

  It is also important that the pilot can adjust their position to obtain the best field of vision and enable quick responses for movement of the various controls. For this reason, the pilot must be able to alter the height and position of their seat to ensure that the controls are in comfortable reach. The temperature, ventilation and lighting in the cockpit must also be adequate and these must be adjustable to suit the individual.

- **CNC Lathe**

  The CNC lathe is computer-operated using a keypad or keyboard. It is important, therefore, to ensure that the operator can access the keypad or keyboard easily and that they can use the keys comfortably. For this reason, the operator must be able to adjust their operating position, i.e. chair height and position, as well as the actual position of the keyboard.

Relationship Between Physical Stressors and Human Reliability

Stress can be caused by a number of factors, including physical stressors, such as extremes of heat, humidity, noise, vibration, poor lighting, restricted workspace, etc. The presence of physical stressors has a negative effect on people and means that errors are more likely to occur.

Physical stressors affect how comfortable a person is and their ability to concentrate and may even make them feel unwell. Different people may be affected by varying degrees of the physical stressor.

For example, some people are not affected by increased room temperature, while others start to feel uncomfortable and may become restless after a few degrees' rise. Pregnant women are more likely to be affected before other members of the workforce. However, if the temperature continued to rise, then more and more people would be affected and the likelihood of errors occurring would rise too as concentration levels dropped. In addition to this, people are more likely to lose their tempers or have decreased levels of patience which, again, may lead to errors or incidents occurring. Eventually, a very warm working environment may result in fainting or heat exhaustion, which could have serious implications in a high-risk environment.

Some environments are very warm by their nature, e.g. working in a busy kitchen. Procedures should be in place to ensure that individuals are protected from excessive heat, e.g. regular rest breaks away from the heat, availability of cold drinks, good air circulation, etc.

In order to prevent errors, or reduce them as far as possible, you need to ensure that the working environment is as comfortable as possible. Where physical stressors are likely to be a problem, e.g. in a noisy environment, other controls must be in place to prevent them affecting an individual's ability to work safely. These controls may be in the form of suitable PPE, limited time within the environment, or regular breaks, for example.

Effects of Under-Stimulation, Fatigue and Stress on Human Reliability

As we have said, **stress** is the reaction that people have to excessive pressure and occurs when they worry that they can't cope. Stress can affect performance and an individual's ability to make decisions and work effectively.

Both work overload (having too much to do or the work being too difficult) and work underload (routine, boring and under-stimulating tasks) can be sources of stress:

- **Under-stimulation**

  With advances in technology, jobs can become more monotonous and controlled if they are designed to minimise skill requirements, maximise management control and minimise the time required to perform a task. Such jobs are likely to create negative attitudes and poor mental and physical health. It is only through re-designing such work that improvements can be made in the quality of working life and the performance on the job.

- **Fatigue**

  Fatigue can be defined as 'weariness after exertion' or can occur after repeated periods of stress. Severe fatigue can lead to poorer performance on tasks requiring attention, decision-making or high levels of skill. Shift work, working at night or extended hours can all result in fatigue and have an adverse effect upon health. For safety-critical work, such as train driving, the effects of fatigue can give rise to increased risks.

  Shift work, especially night-working, can impact on safety. During the night, job performance may be poor and tasks completed more slowly. The hours between 02.00 and 05.00 are the highest risk for fatigue-related conditions. Sleep loss can lead to lowered levels of alertness. Sleep debt, which is a build-up of sleep loss, leads to reduced levels of productivity and attention. These effects can also affect early morning shift workers and people who are on call.

> **MORE...**
>
> You can find further information on the effects of fatigue on human performance at:
>
> www.hse.gov.uk/ humanfactors/topics/ fatigue.htm

- **Stress**

  The introduction of new systems can also be a source of stress where complicated technology and the absence of training and support can exert undue pressure on individuals. There are also factors intrinsic to the job that can act as stressors, such as:

  - Poor physical working conditions (e.g. high levels of noise, poor ventilation).

  - Working inconvenient and excessive hours.

  - Working on a repetitive and fast-paced task.

  - Having a job which involves risk or danger.

> **DEFINITION**
>
> **STRESS**
>
> The reaction that people have to excessive pressure or other types of demands placed on them.

When attempting to improve job satisfaction and reduce stress levels, organisations often focus on the individual worker by providing stress management courses and employee assistance programmes. These are attempts to deal with the problem on an individual basis, whereas the longer-term solution is to consider organisational and job design issues in order to deal with the underlying work-related causes.

STUDY QUESTIONS

9. What effects might shift work have on an individual's performance?

10. How might the system of payment or terms of employment at work affect an individual's performance?

11. Explain the term 'ergonomics' and discuss how the poor application of ergonomics might lead to injury and occupational ill health.

12. What features are present in an ergonomically-designed crane cab control system?

(Suggested Answers are at the end.)

Behavioural Change Programmes

IN THIS SECTION...

- Behavioural change programmes aim to change individual behaviour by positively reinforcing desired behaviour and deterring undesired behaviour.

- They rely on:

  - Observations by supervisors and other workers.

  - Providing prompt feedback to improve behaviour.

Principles of Behavioural Change Programmes

Behavioural programmes aim to change individual behaviour using a number of techniques including:

- Observations.

- Feedback.

- Goal-setting.

- Team-working.

There are many different types of behavioural change programmes available, but all are based on the same fundamental principles.

The key principle is to positively reinforce the desired behaviour and deter or even punish the undesired behaviour.

The first step is to identify the desired behaviour. The behaviour should be specific, observable and easily measured. In fact, simply observing behaviour can, in itself, lead to a positive improvement in the behaviour, but this will only be a temporary effect.

Behavioural programmes can change individual behaviour

TOPIC FOCUS

Steps of a behavioural change programme:

Step 1: Identify the specific observable behaviour that needs changing, e.g. increased wearing of hearing protectors in a high-noise environment.

Step 2: Measure the level of the desired behaviour by observation.

Step 3: Identify the cues (or triggers) that cause the behaviour and the consequences (or pay offs) (good and bad) that may result from the behaviour.

Step 4: Train workers to observe and record the safety critical behaviour.

Step 5: Praise/reward safe behaviour and challenge unsafe behaviour.

Step 6: Feed back safe/unsafe behaviour levels regularly to workforce.

Such a change programme is best illustrated with an example – speeding in a car.

What are the triggers that may cause somebody to speed?

These could be:

- Late for an appointment.
- Emergency.
- Road rage.
- Listening to exhilarating music.
- Empty road.

What about the consequences? These can be either rewards or punishments:

- Arrive early/on time – reward.
- Feel good – reward.
- Have an accident – punishment.
- Stopped by police – punishment.
- Increased wear and tear – punishment.

A consequence will have a much greater impact if it:

- Happens sooner rather than later after the behaviour.
- Is certain to happen rather than unlikely.
- Is important to the individual.

So, with our driving example, if every time we exceeded the speed limit we were immediately fined, motorists would soon adhere to the required limits.

It is important to focus on the safe/unsafe behaviours and not the desired outcome of the programme. So, if the objective of such a programme is to reduce the incident/accident rate, then it should focus on safe behaviours (and reward those) and not on the injury rate. If we rewarded a low injury rate, then this would encourage under-reporting rather than safe behaviours.

Many behavioural change programmes identify a few key behaviours that have, perhaps, led to accidents previously, or gave cause for concern – for example, the failure to wear gloves when handling knives, or to wear a seatbelt while on a forklift truck. The desired behaviours are then identified (such as 'when driving the forklift truck the operator wears his seatbelt') and the observers then observe against this specific behaviour. If the operator is wearing his seatbelt, then positive reinforcement and thanks are given. If the operator is **not** wearing his seatbelt, then the observer highlights their concern and discusses any barriers to safety that could have resulted in the action. By being observed and given feedback regularly, workers' behaviour changes, such that the correct behaviours become almost good habits. At this point, a new behaviour may be added to the observation sheet, and the process continues.

Organisational Conditions Needed for Success in Behavioural Change Programmes

Behavioural programmes should not be viewed as quick fixes and, unless they are properly resourced with overt continuing management commitment, they are unlikely to succeed.

As the behavioural change process hinges on the ability to give and receive positive and negative feedback, the organisation and its employees must be ready to accept this step. It may seem strange to have someone thanking you for wearing PPE, but this positive feedback is a critical part of the process. Equally, any negative feedback is there for guidance and so should also be received with an open mind, which can be a struggle if the safety culture is not well developed.

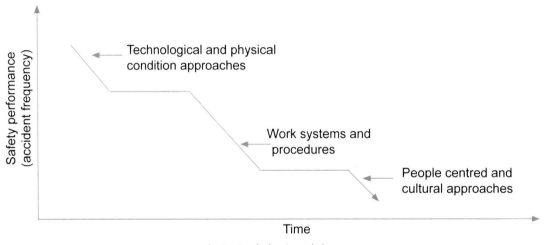

Actioning behavioural change

This graph illustrates that, typically, organisations first consider the technical issues that affect safety, such as having safe equipment and premises. When these are in place, they then turn to ensuring that the systems of work and procedures are satisfactory. Unless these two approaches are in place, a behavioural programme is unlikely to work.

A behavioural programme is more likely to be effective if the reward/punishment is:

• Likely or even certain.

• Important to the individual.

• Given soon after the safe/unsafe act.

Feedback needs to be provided very soon after the safe/unsafe act so that the safe behaviour is reinforced, not only to the individual but to all those affected, so that they appreciate the impact of the programme, e.g. collective results published weekly.

In one published study, workers were provided with earplugs to protect them from very high noise levels. The initial usage rate was only 35%. After a two-month programme in which the wearing of the plugs was rewarded with tokens, the usage rate had increased to 90%. The scheme was finished and it was found that usage had been maintained a further three months later. The initial discomfort often experienced by wearers of hearing protection had worn off and when users removed the ear plugs, their heightened awareness of the high noise levels further reinforced the desired behaviour.

Example of the Content of Typical Behavioural Change Programmes

All programmes need behaviour which can be easily observed and assessed.

In a factory, a process involves loading a pallet with 25kg sacks of cement and then transporting the pallet to a lorry for despatch.

A checklist is then developed to identify the expected behaviour and record the number of safe and unsafe acts. Here is an extract:

Task	Expected Behaviour	Safe	Unsafe	Not Seen	Comment
Loading pallet	Loader wears safety gloves	3	21		
	Loader wears safety shoes	3	2		
	Loader adopts safe lifting procedure	2	0		
	Loader keeps environment tidy	1	1		
	Loader deals with spillages	0	0	1	
Transporting pallet by FLT	Driver sounds horn when approaching exit doors	4	1		
	Driver keeps within speed limit	4	2		
	Driver keeps forks lowered	2	0		
	Driver is courteous	2	0		
TOTAL		21	7		
% safe		75%			

FLT = forklift truck

For each of the two tasks, a list of expected observable behaviours is identified. Observers then regularly visit the workplace and observe the behaviour and record whether it was safe, unsafe or not seen. Observers may include all workers and should not be just those with management or supervisory roles.

Following each observation, the feedback is given soon afterwards, either individually or as a team, in which safe behaviours are praised and unsafe behaviours discussed. The worker(s) observed are invited to give feedback and to explain, for example, why it was not feasible to wear gloves. The discussion may lead to suggestions as to how to change the task to improve safety.

Clearly, the discussion will need to be handled carefully and should not create hostility.

Following a series of observations, the percentage of unsafe behaviour can be calculated and publicised:

$$\text{Percentage safe behaviour} = \frac{\text{Sum of safe observations}}{\text{Sum of safe + sum of unsafe behaviours}} \times 100$$

STUDY QUESTION

13. Outline the steps of a behavioural change programme.

(Suggested Answer is at the end.)

Summary

Human Psychology, Sociology and Behaviour

Factors that influence human behaviour include:

- Personality – how extrovert or introvert a person is.

- Attitude – their beliefs about safety.

- Aptitude – ability.

- Motivation – what inspires them to work safely.

There are a number of **theories of human motivation**:

- **Mayo (Hawthorne Experiments)** – the simple fact that workers were the subject of experimental study improved their performance.

- **Maslow (hierarchy of needs)** suggested five levels of need: biological, safety or security, social, esteem and self-actualisation.

- **Vroom's expectancy theory** proposes that choices made by the individual are based on estimates of how well the expected results of a given behaviour will lead to the desired results.

- **Blanchard model of motivational outlooks** – people have a natural desire to grow, develop, and do meaningful work and the key psychological needs of an individual leading to motivation are autonomy, relatedness and competence.

Experience, social background and education/training affect behaviour at work.

Perception of Risk

In relation to **human sensory receptors**:

- Each of our senses works in the same way by sending signals to the brain.

- We tend to screen out things we are not interested in.

- Sensory defects increase with age and ill health.

When perceiving danger:

- **Perceptual set** is dangerous because we assume both the danger and the solution without seeing the real issues.

- Our perception of hazards can be **distorted**.

- **Errors of perception** can be caused by physical factors, such as fatigue and stress.

Human Failure Classification

HSG48 identifies two types of human failure: errors (accidental) and violations (deliberate).

Errors are actions or decisions which were not intended, involved a deviation from an accepted standard, and which led to an undesirable outcome.

Errors can be characterised as: slips, lapses and mistakes.

There are three types of violation: routine, situational and exceptional.

Rasmussen's model of skill-, rule- and knowledge-based behaviour states that:

- **Skill-based behaviour** describes a situation where a person is carrying out an operation without the need for any conscious thought.

- **Rule-based behaviour** is at the next level – a situation where the operator has rules which they can apply to deal with a specific situation.

- **Knowledge-based behaviour** is for situations where there are no tried rules or routines or the necessary skills. Trial and error may be the only method available.

In many **major disasters**, such as the *Herald of Free Enterprise*, human error has been shown as a major contributory factor.

Improving Individual Human Reliability in the Workplace

Motivation may be improved by:

- Workplace incentive schemes.

- Reward schemes.

- Job satisfaction.

- Appraisal schemes.

- Selection of individuals by:

    - Matching skills and aptitudes.

    - Training and competence assessment.

    - Fitness for work schemes and health surveillance, if appropriate.

    - Support for ill health and stress-related illness.

Organisational Factors

Weaknesses in the safety management system increase the probability of human failure. These include failures in:

- Policy.

- Planning.

- Setting of standards.

- Information.

- Responsibilities.

- Monitoring.

Groups, both formal and informal, within an organisation affect the control of risks.

Communication mechanisms within an organisation vary in their complexity, reliability and formality.

Communication between and within groups is important. It can be:

- Horizontal or vertical.

- Inward and outward.

Job Factors

The way in which work is organised for individuals with respect to **shift patterns**, **means of payment** and **patterns of employment** can have an important effect on the way they carry out their job.

Task analysis is a process that identifies and examines tasks performed by humans as they interact with systems. By breaking the task down into each step, the cause of an injury may become apparent, and the analysis may identify a better way of completing the task.

The **design** of the work environment can have an effect on human reliability. The following are some issues to consider:

- Displays should be arranged so they can be scanned with minimum effort.

- Consistency is important in the action of control devices.

- Factors such as noise, dust, smell, vibration, temperature, lighting levels and humidity all contribute to a worker's ability to concentrate.

The **ergonomist's** skills include:

- **Anthropometry** – a study of human measurements, such as shape, size, and range of joint movements.

- **Physiology** – a study of the calorific requirements of work (how much energy is needed) and body functions, the reception of stimuli, processing and response.

Human performance can deteriorate due to **poor design** of workstations, such as those that are too low, or too high. Work that involves repetitive movements can lead to **upper limb disorders**.

Physical stressors affect how comfortable a person is and their ability to concentrate and may even make them feel unwell. These include: extremes of heat, humidity, noise, vibration, poor lighting, restricted workspace, etc.

Fatigue can be defined as 'weariness after exertion' or can occur after repeated periods of stress. Severe fatigue can lead to poorer performance on tasks requiring attention, decision-making or high levels of skill.

Behavioural Change Programmes

Behavioural change programmes endeavour to change individual worker behaviour by positively reinforcing desired behaviour and deterring undesired behaviour:

- Specific observable behaviour to be changed is identified and then measured.

- The cues that encourage the behaviour and resulting consequences are identified.

- Safe behaviour is encouraged/rewarded.

- Unsafe behaviour is challenged.

Such programmes rely on:

- Observations by supervisors and other workers.

- Providing prompt feedback to improve behaviour.

Exam Skills

Here, you can have another attempt at a 20-mark, Section B question; this one is on the topic of human error. Remember you would allow yourself 30 minutes to answer this question in the exam.

QUESTION

Outline the desirable design features of controls **AND** displays on a control panel for a complex industrial process aimed at reducing the likelihood of human error. **(20)**

Suggested Answer Outline

Remember that there are 20 marks available, so try to mention 22 points in outline to maximise the chance of getting full marks.

With this question on ergonomics, you need to focus on features of controls and displays that would reduce the rate of human error, including examples and reasons in your answer.

The examiner would be looking for an answer including points similar to the following:

- Controls – keep to a minimum, while still ensuring safe operation and control.

- Controls should have a positive action.

- The user requires immediate feedback.

- Stop functions should:

  - Be easy to activate.

  - Have clear markings and functions.

  - Be designed to avoid accidental operation.

  - Be positioned away from interference by non-authorised users.

  - Avoid non-intentional re-start or shut down.

- Controls should be visible and in a logical order such that operators interact with them – up/down buttons, not side by side with each other.

- Select the correct type of control for the operation desired – levers versus knobs.

- Follow colour-coding standards: green – on; red – emergency stop, etc.

- Controls located adjacent to the displays, so the operator can see the consequences of their actions.

- Clearly labelled.

- Language/terminology appropriate for understanding and recognition.

- Displays should attract a response when required: flashing, noise, etc.

- Correct type of display – analogue versus digital.

- Displays should indicate normal and abnormal situations.

- Environmental concerns – located away from heat, glare or anything else that may detract from recognition of any signal.

- Displays should be located an appropriate distance away from the operator for ease of reading/correctly sized display.

- Discussion/consultation with users to obtain feedback on suitable design and operation to avoid confusion.

Example of How the Question Could Be Answered

For a control panel and controls to be used effectively to manage the process and ensure safety, they need to be designed from the operator's point of view; this design should focus on the ergonomic layout of the panel and controls, such that they are easy to reach, move and manipulate. The numbers of controls or interactions should be kept to a minimum to prevent mental overload or having too much to monitor/review.

Controls that are safety critical, such as emergency stop buttons, should be clearly identified and be easy to operate from any position. The emergency stop function should override all other control functions. Other start and stop controls should be designed differently, with start buttons recessed in order to prevent accidental operation. Controls should be labelled in the correct language for the users, and designed to accepted cultural norms, e.g. red for 'stop', green for 'go', clockwise to increase, etc. It should also be clear to the user that the operation of the control has been recognised by the process by providing feedback to the user.

The layout of the panel should be logical and follow the operator's sequence if possible, and the right type of control for operation – dials for rotating things rather than buttons to push.

Other things that need to be considered would involve the environmental conditions in which the panel will be sited and operated, light, fumes, indoors/outdoors, etc.; the controls need to be suitable for the environment (buttons fading in sunlight).

Alarm signals must be clear and obvious and give operators time to react to the condition as well as being relevant to the emergency, so they don't become blasé to them going off and just mute them and forget. An alarm to warn of a change of condition should therefore be different to that signalling an emergency situation.

You would also need to think about the operator from the point of view of markings for eyesight, too small so unable to read, sticky labels that may fall off and be put on the wrong way round, colours – use of standards.

When considering the design of control panel displays, it is essential that safety critical information is located in a prominent position where it is not confused with other process information. The display should also be designed using cultural norms and so that the status of the process is clear, e.g. by using green lights to show equipment that is 'on' and red to show equipment is 'off'.

The display should follow the process flow sequence so that the operator can relate the information on the control display to the controls that they operate and the process that they understand.

The display must be clear and easy to read, e.g. free from flicker, glare and reflections. The use of process diagrams rather than text may also aid understanding.

Finally, alarm conditions on displays should be clear and prominent – it may be desirable to have a 'warning' alarm sequence and also a different 'critical' alarm utilising beacons and audible alarms, making it very clear that emergency action must be taken.

Reasons for Poor Marks Achieved by Candidates in Exam

An exam candidate answering this question would achieve **poor marks** if they made the mistake of describing the features of a control room or panel, rather than the design features of controls and control panels.

The Role of the Health and Safety Practitioner

Learning Outcomes

Once you've read this element, you'll understand how to:

1 Explain the role of the health and safety practitioner.

2 Explain the importance of effective communication and negotiation skills when promoting health and safety.

3 Outline how health and safety practitioners can use financial justification to aid decision-making.

Contents

Role of the Health and Safety Practitioner

IN THIS SECTION...

- The health and safety practitioner must be aware of the significance of their role in:
    - Protecting employees, employers and third parties from risk or injury in the workplace.
    - Evaluating and developing their own competence, and mentoring and supporting the development of health and safety competence in others.
    - Developing, implementing, maintaining and evaluating health and safety management systems.
    - Contributing to the organisation achieving its objectives.
    - Applying ethical principles when working in the health and safety profession.
- Sensible risk management involves organisations taking responsibility for their risks and proportionate steps to manage those risks.
- An organisation's risk profile assesses the likelihood of adverse effects occurring, the associated levels of disruption and costs, and the effectiveness of control measures in place.

NOTE...

Please note that Element A11 will **NOT** be examined in the Unit A exam, but only assessed as part of the Unit DNI assessment.

Role of the Health and Safety Practitioner

The role of the health and safety practitioner can be seen as seeking to minimise the risk of harm or injury at work by educating colleagues, setting procedures and building a culture of safety in the workplace.

This may involve:

- Advising managers on how to comply with health and safety legislation.
- Drawing up strategies, policies and procedures for the organisation.
- Carrying out site visits and safety audits, and identifying potential hazards.
- Designing and delivering training courses on health and safety issues.
- Investigating accidents and finding ways to prevent them happening again.
- Advising on specialist areas such as fire precautions and control of hazardous substances.
- Producing written reports.
- Keeping records, including accident statistics.

A health and safety practitioner should minimise the risk of harm or injury at work

Health and safety practitioners need to work closely with managers, employees and sometimes trade union representatives. They may also liaise with external contacts, such as contractors, clients and the Health and Safety Executive (HSE).

> **TOPIC FOCUS**
>
> The skills and personal qualities required by a health and safety practitioner include:
>
> - Strong interpersonal skills, for negotiating changes in the workplace and influencing others to adopt them.
>
> - An understanding of health and safety legislation and how to interpret it.
>
> - A sound knowledge of technical and operational processes.
>
> - The ability to think ahead and anticipate potential problems.
>
> - Presentation skills, in order to lead training sessions.
>
> - An ability to communicate complex information in a straightforward way.
>
> - A clear writing style.
>
> - A thorough and methodical approach.
>
> - Persistence, patience, adaptability and some degree of physical fitness if working outdoors or in major plants.

To do this properly, health and safety practitioners need to:

- Be properly trained and suitably qualified.

- Maintain adequate information systems on topics including civil and criminal law, health and safety management and technical advances.

- Interpret the law in the context of their own organisation.

- Be involved in establishing organisational arrangements, systems and risk control standards relating to hardware and human performance, by advising line management on matters such as legal and technical standards.

- Establish and maintain procedures for reporting, investigating, recording and analysing accidents and incidents.

- Establish and maintain procedures, including monitoring and other means, such as review and auditing, to ensure senior managers get a true picture of how well health and safety is being managed (where a benchmarking role may be especially valuable).

- Present their advice independently and effectively.

In terms of organisational structure/relationships, health and safety specialists must:

- Support the provision of authoritative and independent advice.

- Have a direct reporting line to directors on matters of policy, and the authority to stop work if it contravenes agreed standards and puts people at risk of injury.

- Have responsibility for professional standards and systems; on large sites or in a group of companies, they may also have line management responsibility for other health and safety specialists.

With regard to relationships outside the company, they must liaise with a wide range of bodies, including:

- Local authority environmental health officers and licensing officials.

- Consultants/contractors.

- Fire service.

- Insurance companies.

- HSE.

Potential Conflicts

Responsibilities for health and safety are owed by a number of parties in the workplace. The **employer** has responsibilities for employees and third parties; **individuals**, who may be employees or third parties, also have legal responsibilities, and **trade union representatives** or **representatives of workplace safety** have certain legal entitlements. The health and safety practitioner needs to work with all these parties in order to maximise health and safety performance in the organisation.

The employer will normally appoint the health and safety practitioner. A key aim will be to protect employees and third parties affected by work activities, and good working relationships with safety representatives will be necessary to achieve effective consultation with and co-operation from the workforce. The health and safety practitioner needs to be able to work impartially with all these different parties.

A conflict of interest may arise when an individual has to make a decision at work that may affect their private interests.

For example, the safety manager of a large organisation has the task of appointing a new safety adviser. One of the candidates for the post is a good friend of the manager, who expects favourable treatment even though they may not be the best qualified and experienced candidate. The conflict of interest dilemma the manager has is whether to let a personal interest interfere with their professional judgment.

What should you do when a conflict of interest arises?

There are two aspects to dealing with conflicts of interest:

*   Identifying and disclosing the conflict of interest, which is primarily the responsibility of the individual who is subject to the conflict. It is clearly better to err on the side of openness even when the situation is not clear-cut, particularly in the long term when the conflict may become more widely known and more difficult to resolve, leading to the possible accusation of bias or even dishonesty.

*   Deciding what action (if any) is necessary to avoid or mitigate any consequences, usually the responsibility of the manager or department in which the conflict has arisen. This may range from taking no action at all to, in extreme circumstances, the resignation or dismissal of the individual concerned.

In this example, the likely outcome if the conflict of interest was declared before the appointment was made would be for the manager not to sit on the appointment panel. If, however, the possible conflict was concealed and later became known, then the manager could be accused of unprofessional conduct. (We will look at the topic of professional conduct later in this element.)

Meaning of the Term 'Competence'

'Competence' is a difficult term to define. In the **Management of Health and Safety at Work Regulations 1999** 'competence' in respect of the competence of a health and safety practitioner is described as having *"sufficient training and experience or knowledge and other qualities to enable him properly to assist in undertaking the measures referred to..."*.

Accordingly, the term reflects a combination of knowledge and relevant experience, although these terms are not mutually exclusive.

The **Electricity at Work Regulations 1989** require *"persons to be competent to prevent danger and injury"* and sheds some light on what competence actually means. Knowledge and experience, along with understanding, is referred to but an additional important requirement is the ability to recognise at all times whether it is safe to continue working. This is relevant to general health and safety practice where the practitioner needs to recognise the limits of their own competence. An apt quote in this respect states: 'The most important thing in terms of your circle of competence is not how large the area of it is, but how well you've defined the perimeter'. So, in the role of health and safety practitioner, it is vital to know the limits of your advisory capacity. An essential ingredient of competence is recognising its limits, i.e. the point at which you begin to get involved in an area which is beyond your competence and you need to either call in external expertise or upgrade your skills.

Need for Health and Safety Practitioners to Evaluate Their Own Practice

As well as competency in practical risk management, the practitioner needs to evaluate and develop their own practice. This is in line with the Institution of Occupational Safety and Health (IOSH) requirement for continuous professional development, and ensures that the practitioner maintains and develops competency, keeps up to date and remains effective.

The broad requirements are that practitioners should:

- Reflect on their own practice.

- Review their practice against appropriate goals.

- Set and prioritise goals and targets for self-improvement.

- Adapt their own practice in the light of changes in professional practice.

In order to achieve this, practitioners need to:

- Identify goals and targets which could be set in a number of ways, such as:

  - From national standards.

  - From assessment of current competency.

  - From anticipated future demands.

  - From personal aspirations.

  - To meet organisational needs.

- Review their own performance, which might involve evaluating work results, undergoing appraisals or formative assessments, or seeking the views of colleagues and clients.

- Develop their personal action plans and monitor their achievement.

- Develop and change their own practice, and evaluate the effectiveness of the developments.

Self-improvement targets need to be set

- Anticipate and identify change, and respond appropriately. This could arise from changes in professional practice, from national and local systems or from changes to organisational policy and procedures.

These requirements can be divided into two principal components: evaluation, and identification of self-development needs.

Evaluating Own Practice

Performance Criteria

The practitioner should be able to:

- Evaluate their own practice against set targets and goals.
- Use a range of valid and reliable evidence to assess their own work, which includes an assessment of behaviour and values by others.
- Involve others in the interpretation of evidence.
- Use evidence to reflect on their own practice and professional issues.
- Accept criticism in a positive manner, and assess its validity and importance.
- Revise goals and targets in the light of their reviewing evidence and performance.

Identifying Self-Development Needs

Performance Criteria

The practitioner should be able to:

- Set and prioritise clear and realistic goals and targets for their own development.
- Base goals and targets on the accurate assessment of all the relevant information relating to their own work and achievement, including developments in professional practice and related areas.
- Devise a personal action plan and review it regularly.
- Try out developments in their own practice in a way which does not cause problems for others.
- Evaluate developments in their own practice and ensure continued self-development.

Health and Safety Practitioner – Mentoring and Supporting

Health and safety practitioners are often involved in training where instruction and demonstration can help managers to understand how to perform tasks such as inspections, accident investigations and risk assessments. However, mentoring goes further than this and aims to enhance an individual's skills, knowledge or work performance through one-to-one discussions. The health and safety practitioner, as an experienced colleague, can use their specialist knowledge and understanding to support the development of others.

If the health and safety practitioner is acting in an advisory role, the ownership for the solution to a problem should rest with other people; the function of the practitioner is therefore one of support. Increasingly, organisations are looking for practitioners who are collaborative, supportive and helpful rather than purely advisory.

In a mentoring role, the aim is an alliance between the mentor and the mentee. The responsibility for finding the solution to the problem remains with the mentee and the mentor is simply there to guide and support the mentee's exploration, so supporting a manager to discover the best, most practicable solution to a health and safety problem can involve mentoring. Practitioners may not appreciate the pressures and constraints that managers operate under so the practitioner needs to empathise with the manager's position. However, practitioners do know about the law, the standards that need to be achieved and what solutions might be feasible, so this is the specialist knowledge and understanding they can bring to the discussion.

For managers to 'own' health and safety, they need to be encouraged and supported; telling and doing is less effective than a more collaborative approach which supports the development of managers so that they become less dependent on the specialist health and safety practitioner.

Distinction Between Leadership and Management

Management can be defined as 'the organisation and co-ordination of the activities of a business in order to achieve defined objectives'. It keeps the organisation functioning and involves planning, budgeting, staffing, defining responsibilities, measuring performance, and solving problems when things go wrong.

Leadership is very different. It is the activity of leading a group of people or an organisation and involves establishing a clear vision, sharing that vision with others and providing the information, knowledge and methods to realise that vision. The workforce needs to be aligned to the vision through buy-in and communication in order to change rather than continue to do what they have always done well.

The following terms indicate the differences between a manager and a leader:

Manager	Leader
Systems	People
Control	Trust
How and when	What and why
Imitates	Originates
Accepts	Challenges
Subordinates	Followers
Maintains	Develops
Does things right	Does the right thing

The role of the health and safety practitioner can align with both the management and the leadership models.

One function of the health and safety practitioner is to look after the health and safety management system and advise on aspects such as:

- Formulating and developing health and safety policies and plans.
- Profiling and assessing risks and organising activities to implement the plans.
- Measuring performance.
- Reviewing performance and taking action on lessons learnt.

In other words, the practitioner can be seen as a manager maintaining systems.

But an equally important function is that of an agent of change. In this role, the health and safety practitioner needs to galvanise the management board into action to:

- Set the direction for effective health and safety management.
- Establish a health and safety policy that is an integral part of the organisation's culture and value.
- Take the lead in ensuring the communication of health and safety duties and benefits throughout the organisation.
- Respond quickly where difficulties arise or new risks are introduced.

In other words, the practitioner should act as a visionary to get board level buy-in for culture change.

Need to Adopt Different Management Styles

Health and safety practitioners work closely with different groups of people, such as managers, employees, trade union representatives, contractors, clients and enforcement officers. Consequently, the management style they adopt needs to match the nature of the interaction and the type of relationship with each party.

Management styles can broadly be categorised into the following three groups and each has relevance to different facets of the health and safety practitioner role:

- **Autocratic**

  Managers make all the important decisions and closely supervise and control workers. They simply give orders (one-way communication) that they expect to be obeyed, and do not consult.

  This approach is effective when quick decisions are needed or when controlling large numbers of low-skilled workers.

  Health and safety practitioners may need to operate in this autocratic role when:

  Autocratic

  - They have the expertise and authority to closely control a workplace activity.

  - The requirements to ensure safety and health are clear and well defined.

  - There is no need for input from the workforce.

- **Democratic**

  Managers trust workers and encourage them to make decisions. They delegate the authority to do this and listen to their advice and feedback. Decisions are made by the group, by consulting or by vote. This style requires effective two-way communication and may involve discussion groups offering suggestions and ideas.

  This approach is effective with motivated individuals who are capable of making their own decisions or when there is no need for central co-ordination. Health and safety practitioners may find this approach in evidence in the health and safety committee, where binding decisions on policy and procedures are made after participation and final agreement by the committee members.

  Democratic

- **Participative**

  Managers are concerned about the needs and views of their workers and how happy they feel. They consult on issues and listen to feedback or opinions but still make the final decision, albeit in the best interests of the workers, believing that they still need direction.

  This approach is effective where team agreement is important but can be difficult to manage where there are many differing opinions.

  Participative

  Health and safety practitioners may find that the participative approach is essential in collaborating with and supporting managers to encourage their ownership of health and safety. Although the practitioner knows about the law, the standards that need to be achieved and what solutions might be feasible, the responsibility for finding the solution to the problem remains with the manager. The health and safety practitioner is there to guide and support the manager to discover the best and most practicable solution to the problem.

Role of the Health and Safety Practitioner on Safety Management Systems

Health and safety practitioners are those likely to be appointed by employers to help them in managing health and safety in the organisation. Health and safety practitioners need to have the status and competence to advise management and employees, or their representatives, with authority and independence. As we have noted above, they are well placed to advise on many aspects of the safety management system, such as:

- Formulating and developing health and safety policies and plans, not just for existing activities but also with respect to new acquisitions or processes.

- Profiling and assessing risks and organising activities to implement the plans.

- Measuring performance by assessing how well the risks are being controlled and investigating the causes of accidents, incidents or near misses.

- Reviewing performance, re-visiting plans, policy documents and risk assessments to see if they need updating and taking action on lessons learnt, including from audit and inspection reports.

Meaning of the Term 'Sensible Risk Management'

"The concept of sensible risk management aims to balance the growing risk-averse attitude of society toward innovation and development. Consequently, taking a sensible approach to risk management involves:

- *ensuring that workers and the public are properly protected*

- *enabling innovation and learning, not stifling them*

- *ensuring that those who create risks manage them responsibly and understand that failure to [do so] is likely to lead to robust action.*

- *providing overall benefit to society by balancing benefits and risks, with a focus on reducing significant risks – both those which arise more often and those with serious consequences*

- *enabling individuals to understand that as well as the right to protection, they also have to exercise responsibility*

It is not about:

- *reducing protection of people from risks that cause real harm*

- *scaring people by exaggerating or publicising trivial risks*

- *stopping important recreational and learning activities for individuals where the risks are managed*

- *creating a totally risk-free society*

- *generating useless paperwork mountains"*

Source: Sensible risk management, Health and Safety Executive (HSE) (www.hse.gov.uk/risk/principles.htm)

Enabling Work Activities as Part of Proportionate Risk Management

Health and safety at work legislation is about reducing death, serious injury and ill health in workplaces – it is about taking the necessary action to reduce significant risks arising from work. However, in an increasingly risk-averse society, it can degenerate into simply banning activities.

The health and safety practitioner should be clear about what is a legal requirement. Rather than attempt to stop or limit an activity, or assume that it will have undesirable or unintended consequences, the practitioner should therefore check whether the decision or the chosen precautions are proportionate by considering the actual risks.

Understanding what the actual risk is involves considering the likelihood and consequences of something going wrong, and this means thinking about:

- What type of incident the decision or precaution is intended to prevent.

- What injury or ill health could be caused.

- How likely it is to happen.

Organisations should take ownership of their risks and take proportionate steps to manage those risks so that attention is focused on the significant risks that cause injury and ill health, and not the trivia or everyday low risks.

Organisational Risk Profiling

MORE...

This section is based on the UK HSE's description of the process of risk profiling. You will find out more at:

www.hse.gov.uk/managing/risk-profiling.htm

www.hse.gov.uk/managing/delivering/do/profiling

www.hse.gov.uk/managing/delivering/key-actions/key-actions-in-effective-risk-profiling.htm

Purpose

The risk profile of an organisation is a key factor in determining the approach that needs to be taken to manage its health and safety risks. In simple terms, the 'riskier' the organisation, the more effort is needed to manage those risks.

Every organisation will have its own risk profile, and knowledge of the nature of the business will quickly conjure up what health and safety issues need to be addressed, i.e. a call centre sited next to a light engineering company in the same business park.

The risk profile is the starting point for determining the greatest health and safety issues for the organisation. In some businesses, the risks will be tangible and immediate safety hazards, whereas in other organisations the risks may be health-related and it may be a long time before the illness becomes apparent:

- The **aim** of risk profiling is to examine the nature and level of the threats faced by an organisation and the likelihood of these adverse effects occurring (i.e. severity and likelihood). This establishes the likely level of disruption and cost associated with each type of risk and enables the effectiveness of controls in place to manage those risks to be assessed.

- The **outcome** of risk profiling is that significant risks have been identified and prioritised for action and minor risks simply noted to be kept under review. It also informs decisions about what risk controls measures are needed.

Risk profiling is a key activity for leaders and line managers so that they know the risks their organisations face, rank them in order of importance and take action to control them.

Practicality

A risk profile examines the nature and levels of threats faced by an organisation. It assesses the likelihood of adverse effects occurring, the level of disruption and costs associated with each type of risk and the effectiveness of the control measures in place.

The practical implementation of risk profiling in an organisation is driven at different levels.

Leaders need to:

- Identify who takes ownership of health and safety risks.

- Consider the consequences of the worst possible occurrence for the organisation.

- Ensure that risk assessments are carried out by a competent person.

- Maintain an overview of the whole process.

- Identify who will be responsible for implementing risk controls and over what timescale.

Managers should:

- Identify their health and safety risks and prioritise them.

- Ensure that risks are owned so that appropriate resources can be allocated.

- Consider groups that might be at increased risk, such as young or inexperienced workers, pregnant workers and workers with a disability.

- Decide whether control measures in place are adequate or if further action is needed.

- Be prepared to implement interim measures to minimise the risks if some control measures are long term.

- Report risk control performance regularly internally and consider whether it should be done externally.

- Make sure paperwork is kept to the minimum levels necessary.

- Review the organisation's risk profile regularly.

Workers need to:

- Understand the organisation's risk profile.

- Have the necessary information, instruction and training to deal with the identified risks.

- Be consulted in all parts of the organisation to ensure that all areas of risk have been identified.

Organisational Context

Risk profiling provides organisations with a detailed picture of the:

- Risks inherent in its operations.

- Effectiveness of the controls in place to mitigate the risk.

- Framework for monitoring its higher risk priorities.

A risk profile examines the level of threat to an organisation

Some organisations may be willing to accept or retain risk where others may seek to implement risk management strategies to reduce or control, transfer or avoid risk (see Element A8).

Every organisation has its own risk profile – in some, it may principally consist of immediate safety hazards, while in others the risks may be predominantly longer-term, health-related issues. In addition, health and safety risks can range from high-hazard, low-frequency events, such as major explosions, to low-hazard, high-frequency occurrences, such as slips and trips.

However, regardless of the specific nature of the workplace activities, the risk profile should include:

- The key strategic and operational health and safety risks.
- Quantification of these risks in terms of likelihood and severity.
- Identification of current controls and degree of effectiveness.
- Plans for required additional controls.

This enables the organisation to:

- Identify and prioritise the significant risks without giving minor risks unnecessary priority.
- Reduce these risks to an acceptable level.
- Minimise the associated paperwork and bureaucracy.

Contribution of the Health and Safety Practitioner in Achieving the Objectives of an Organisation

Health and safety practitioners can contribute to the achievement of the objectives of an organisation by leading on health and safety issues. They can act as advocates, persuading both managers and the workforce of the value of their knowledge and expertise.

Organisational health and safety objectives are achieved through a planning process which considers where the organisation is now and where it needs to be, what it wants to achieve, who will be responsible for what, how it will achieve its aims and how it will measure its success. This needs to be documented in a policy and a plan to deliver it, where the input of the health and safety practitioner will be instrumental. A key component of this is the risk profile of the organisation and the assessment of these risks to decide what needs to be done to manage these risks, what the priorities are and which are the biggest risks. This is where the knowledge and expertise of the health and safety practitioner comes to the fore.

In organising activities to deliver the plan, the health and safety practitioner has a significant role in involving the workforce and communicating in such a way that everyone is clear on what is needed; the practitioner will need to discuss issues and lead in developing positive attitudes and behaviours.

Ethics and the Application of Ethical Principles

Ethics is concerned with moral issues, i.e. the judgments we make and our resulting conduct. Just because an action we might take is legal does not necessarily mean it is ethical. This is best explained by identifying the ethical principles we would expect practitioners to adopt, such as:

- Honesty in dealings with clients, etc.
- Respecting others.
- Professional integrity.
- Personal conflicts of interest.

IOSH has a Code of Conduct which all members are expected to follow. This requires members to:

- Owe a loyalty to the workforce, the community they serve and the environment they affect.
- Abide by relevant legal requirements.
- Give honest opinions.
- Maintain their competence.
- Undertake only those tasks they believe themselves to be competent to deal with.
- Accept professional responsibility for their work.
- Make those who ignore their professional advice aware of the consequences.
- Not bring the professional body into disrepute.
- Not recklessly or maliciously injure the professional reputation or business of others.
- Not behave in a way that may be considered inappropriate.
- Not use their membership or position within the organisation or Institution improperly for commercial or personal gain.
- Avoid conflicts of interest.
- Not disclose information improperly.
- Ensure information that they hold necessary to safeguard the health and safety of others is made available on request.
- Comply with data protection principles and relevant legislation.
- Maintain financial propriety with clients and employers and where appropriate be covered by professional indemnity insurance.
- Act within the law and notify the Institution if convicted of a criminal offence.

Similarly, the International Institute of Risk and Safety Management (IIRSM) has a Code of Ethics which requires members to:

- Only advise on or undertake tasks where they are competent to do so.
- Ensure professional competence is maintained and developed.
- Avoid conflicts of interest.
- Inform the appropriate authority of any illegal or unethical, safety-related behaviour.
- Conduct themselves with fairness when dealing with others and not engage in discrimination.
- Act as the faithful agent of their clients or employers and accept responsibility for their own work.
- Assist colleagues in their professional development and support them in following this Code.
- Not bring the Institute into disrepute.

STUDY QUESTIONS

1. Explain the concept of sensible risk management.
2. Identify five examples of how a safety practitioner would be expected to adhere to ethical principles.

(Suggested Answers are at the end.)

Effective Communication and Negotiation Skills

IN THIS SECTION...

A health and safety practitioner needs to use effective communication and negotiation when promoting health and safety to other people in order to:

- Develop an organisation's agreed health and safety objectives.

- Influence ownership of health and safety within an organisation by means of participation, management accountability, consultation and feedback.

- Receive and act on appropriate feedback on health and safety performance.

- Disseminate the health and safety message.

- Ensure that all workers are clear about their roles and responsibilities.

Effective Communication

Communication can be defined as the process of delivering information from a sender to a recipient. To be truly effective, the correct information has to be transmitted, received and understood. Consequently, communication can be a problem in organisations generally, not just relating to health and safety issues. Communication within an organisation is often seen as the single most important area requiring improvement. The messages senior management want to communicate are often not the ones that employees end up receiving.

Two important health and safety messages that need to be communicated are:

Communication can be problematic

- Evidence of clear, visible leadership.

- A common appreciation of how and why the organisation is trying to improve performance.

The health and safety practitioner can facilitate the delivery of both these messages.

Employers need to set up clear lines of communication in order to build up levels of trust among employees which will help to improve morale. Poor communication in the workplace inevitably leads to demotivated staff. Managers need to establish clear, achievable goals for both teams and individuals, outlining exactly what is required and ensuring that all staff are aware of the health and safety objectives at each level in the organisation.

Effective communication about health and safety relies on information:

- Coming **into** the organisation.

- Flowing **within** the organisation.

- Going **out** from the organisation.

The health and safety practitioner can play a key role in ensuring that information into the organisation enables it to monitor:

- Legal developments to ensure they can comply with the law.

- Technical developments relevant to risk control.

- Developments in health and safety management practice.

If the health and safety policy is to be understood and consistently implemented, the following key information needs to be communicated effectively:

- The meaning and purpose of the policy.
- The vision, values and beliefs which underlie it.
- The commitment of senior management to its implementation.
- Plans, standards, procedures and systems relating to implementation and measurement of performance.
- Factual information to help secure the involvement and commitment of employees.
- Comments and ideas for improvement.
- Performance reports.
- Lessons learnt from accidents and other incidents.

Organisations may need to pass health and safety information to others as a legal requirement and this could include:

- Accident or ill-health information to enforcing authorities.
- Information about the safety of articles and substances supplied for use at work.
- Emergency planning information.

Need for Consultation and Negotiation

Two of the key organisational requirements for developing and maintaining a positive health and safety culture are co-operation and communication, and both of these involve consultation.

In this respect, you should note some of the observations of the Robens Report as mentioned in Element A9 (which led to the **Health and Safety at Work, etc. Act 1974 (HSWA)**.

The key benefits from consultation and negotiation are:

- Better employment relations between workers and employers.
- Workers feel more involved and are more likely to co-operate with their employer.
- A safer and less stressful environment is created which contributes to a good safety culture.

It is a legal requirement for employees to be consulted about those health and safety issues in the workplace that affect them. Where trade unions are recognised, consultation must take place through the safety representatives they appoint under the **Safety Representatives and Safety Committees Regulations 1977**. All other employees not represented in this way must be consulted, either directly or by means of representatives elected by those employees that they represent, under the **Health and Safety (Consultation with Employees) Regulations 1996**.

However, successful organisations often go further than strictly required by law and actively encourage and support consultation in different ways. Safety representatives are trained which, in common with all employees, enables them to make an informed contribution on health and safety issues. They are also closely involved in directing the health and safety effort through the issues discussed at health and safety committees. Effective consultative bodies are involved in planning, measuring and reviewing performance as well as in their more traditional reactive role of considering the results of accident, ill-health and incident investigations and other concerns of the moment.

Employees at all levels are also involved individually or in groups in a range of activities. They may, for example, help devise operating systems, procedures and instructions for risk control and help in monitoring and auditing. Supervisors and others with direct knowledge of how work is done can make an important contribution to the preparation of procedures which will work in practice. Other examples of good co-operation include forming ad hoc problem-solving teams from different parts of the organisation to help solve specific problems – such as issues arising from an accident or a case of ill health. Such initiatives are supported by management and there is access to advice from health and safety specialists.

Opportunities to promote involvement also arise through the use of hazard report books, suggestion schemes or safety circles (similar to quality circles) where health and safety problems can be identified and solved. These, too, can develop enthusiasm and draw on worker expertise.

Influencing Ownership of Health and Safety

At board level, responsibility and ownership of health and safety can be achieved by ensuring that:

- Health and safety arrangements are adequately resourced.

- Competent health and safety advice is available.

- Risk assessments are carried out.

- Employees or their representatives are involved in decisions that affect their health and safety.

The board should consider the health and safety implications of change (new processes, working practices, personnel) and allocate adequate resources to the task.

The organisation's health and safety management system should be 'owned' at each level, with everyone who works at the company owning part of it. Everyone has a responsibility to contribute to the system and those who have assigned or designated responsibilities should be accountable to the management or staff of the company for safety performance in their areas of responsibility.

The following activities can act as drivers to influence ownership:

- **Participation** by employees supports risk control by encouraging their 'ownership' of health and safety policies. It establishes an understanding that the organisation as a whole, and people working in it, benefit from good health and safety performance. Pooling knowledge and experience through participation, commitment and involvement means that health and safety really becomes 'everybody's business'.

- **Management accountability** means that managers are judged on how well or effectively they carry out the duties they are responsible for. Consequently, their credibility relies on taking on board their responsibilities and accounting for or explaining the actions taken (or not taken) to senior managers.

- **Consultation** can occur at each stage of the health and safety management system (Plan, Do, Check, Act):

    - Plan – consult workers or their representatives during the planning and organisation of training.

    - Do – involve and consult workers and representatives during the implementation process, by ensuring there are systems in place that allow workers to raise concerns and make suggestions.

    - Check – involve the workforce in setting and monitoring performance measures and encourage them to monitor their own work area.

    - Act – discuss plans for review with workers or their representatives, use information from safety representatives' inspections to feed into the review and discuss the review findings with workers or their representatives.

- **Feedback** on success and failure is an essential element in motivating employees to maintain and improve performance. Successful organisations emphasise positive reinforcement and concentrate on encouraging progress on those indicators which demonstrate improvements in risk control. This also encourages identification with and ownership of the health and safety programme.

Importance of Receiving and Acting on Feedback

Reviewing is the process of making judgments about the adequacy of health and safety performance. Organisations need to have feedback to see if the health and safety management system is working effectively as designed. Reviewing also gives the opportunity to celebrate and promote health and safety successes. Increasingly, third parties are requiring partner organisations to report health and safety performance publicly.

The main sources of information on health and safety performance feedback come from measuring activities and from audits.

Feeding information on success and failure back into the system can help to motivate employees, and successful organisations emphasise positive reinforcement and concentrate on encouraging progress on those indicators that demonstrate improvements in risk control.

Reviewing should be a continuous process undertaken at different levels within the organisation and involving all stakeholders.

Review plans may include:

- Monthly reviews of individuals, supervisors or sections.

- Three-monthly reviews of departments.

- Annual reviews of sites or of the organisation as a whole.

Organisations should decide on the frequency of the reviews at each level and devise reviewing activities to suit the measuring and auditing activities. In all reviewing activity, the result should be specific remedial actions which:

- establish who is responsible for implementation; and

- set deadlines for completion.

These actions form the basis of effective follow-up, which should be closely monitored.

Key performance indicators for reviewing overall health and safety performance might include:

- assessment of the degree of compliance with health and safety system requirements;

- identification of areas where the health and safety system is absent or inadequate (those areas where further action is necessary to develop the total health and safety management system);

- assessment of the achievement of specific objectives and plans; and

- accident, ill-health and incident data accompanied by analysis of both the immediate and underlying causes, trends and common features.

These indicators are consistent with the development of a positive health and safety culture and emphasise achievement and success rather than merely measuring failure by looking only at accident data.

Organisations may also 'benchmark' their performance against other organisations by comparing:

- Accident rates with organisations in the same industry that use similar business processes and experience similar risks.

- Management practices and techniques with organisations in any industry to provide a different perspective and new insights on health and safety management systems.

As part of a demonstration of corporate responsibility, more organisations are mentioning health and safety performance in their published annual reports.

Different Methods of Communication

To communicate and effectively promote the health and safety message, there are three basic methods used. Electronic and social media are now able to incorporate some of these together.

Verbal Communication

This is communication using the spoken word, e.g. face-to-face conversations, meetings, interviews, training sessions, by telephone or over a Public Address (PA) system.

It is the easiest and most commonly used method of communication but there are weaknesses associated with it. If verbal communication is to be used to convey safety critical information to workers, these weaknesses must be overcome.

Limitations	Merits
Language barrier may exist.	Personal.
Jargon may not be understood.	Quick.
Strong accent or dialect may interfere.	Direct.
Background noise may interfere.	Allows for checking of understanding.
Recipient may have poor hearing.	Allows for feedback to be given.
Message may be ambiguous.	Allows for exchange of views.
Recipient may miss information.	Usually allows for additional information to be transmitted by means of tone of voice, facial expression and body language.
Recipient may forget information.	
No written record as proof.	
Poor transmission quality if by telephone or PA system.	

Written Communication

Here communication is by means of the written word, e.g. reports, memos, e-mails, notices, company handbooks, policy documents, operating instructions, risk assessments, minutes of meetings, etc.

Limitations	Merits
Indirect.	Permanent record.
Takes time to write.	Can be referred back to.
May contain jargon and abbreviations.	Can be written very carefully to avoid use of jargon, abbreviations and ambiguity.
Can be impersonal.	Can be distributed to a wide audience relatively cheaply.
Message may be ambiguous.	
Message may not be read by recipient.	
Language barrier may exist.	
Recipient may not be able to read.	
Immediate feedback is not available.	
Questions cannot be asked.	
Recipient may have impaired vision.	

Graphic Communication

Communication can take place by using pictures, symbols or pictograms, e.g. safety signs such as a fire exit sign, hazard-warning symbols such as the skull and crossbones on the label of a toxic chemical, or photographs such as in the operating instructions for a machine showing a guard being used correctly.

Limitations	Merits
• Can only convey simple messages.	• Eye-catching.
• Might be expensive to buy or produce.	• Visual.
• May not be looked at.	• Quick to interpret.
• Symbols or pictograms may be unknown to the recipient.	• No language barrier.
• No immediate feedback available.	• Jargon-free.
• Questions cannot be asked.	• Conveys a message to a wide audience.
• Recipient may have impaired vision.	

Graphic communication

Broadcasting Methods

Verbal, written and graphic communication methods can be used in various ways to broadcast health and safety information. All these broadcasting techniques have strengths and limitations so a mix of some or all of these techniques is usually used to ensure that essential messages are transmitted and correctly understood by all staff.

Notice boards should be 'eye-catching' and located in areas used by all workers, e.g. rest rooms or central corridors. Notices should be current, relevant and tidily displayed. Cluttered, out-of-date, irrelevant notices obscure the messages being conveyed.

Displaying a notice does not mean that it will be read. Typical contents might include: the safety policy; employer's liability insurance certificate; emergency procedures; identity of safety representatives and first aiders; minutes of safety committee meeting; accident statistics, etc.

Noticeboard

Posters and videos are used to provide safety information, drawing attention to particular issues and supporting the safety culture.

Films or videos are mainly used in training programmes and, if well made, can hold the audience's attention.

Procedures for Resolving Conflict and Introducing Change

The introduction of change may be accompanied by conflict within an organisation; it is vital that conflicts are resolved to ensure a good working environment and atmosphere.

During periods of change, conflict can occur because of:

- **Personality clashes**: change bringing people of differing personalities into new relationships.

- **Poor communication**: can lead to misunderstandings and confusion which can fuel conflict.

- **Conflicting interests**: change can alter the power of relationships within an organisation.

- **Lack of leadership and control**: resulting in a lack of clear direction which can lead to conflict as different people interpret the scenario for change in different ways.

> **DEFINITION**
>
> **ORGANISATIONAL CONFLICT**
>
> Any perceived clash of interests between individuals, groups or levels of authority in an organisation.

While tackling the above areas will help to resolve conflict, note that there are two broad approaches to conflict:

- **Unitary Approach**

  This involves the idea of the common aims of the organisation, i.e. its well-being and how workers and management have the same basic interest in that well-being. According to this view, conflicts arise because workers do not fully appreciate where their true interests lie. There is also some blame on management when conflict occurs because management must have failed to communicate with workers and convince them that their best interests lie in co-operation and not conflict. According to the unitary approach, the best way to tackle conflict at its roots is to generate team spirit, company loyalty, and good working conditions.

- **Pluralist Approach**

  This recognises that the organisation is made up of various groups whose interests and goals may differ. Conflicting parties will benefit from identifying issues of compatibility.

  Conflict should be controlled by balancing the various groups. Where strong management works alongside strong trade unions, each side respects the other and does not lightly enter into conflict. The causes of conflict are brought out into the open and hard bargaining takes place, but serious disruption to the work of the organisation is avoided.

Generally, managers take the unitary approach to conflict and change, while trade unions favour the pluralist approach.

Ensuring Roles and Responsibilities Are Clear, Understood and Implemented by Workers

The promotion of ownership of health and safety in an organisation is an important aspect of the health and safety practitioner's role. Ownership relies on all workers understanding what their personal role is within the health and safety management system and this should be clearly defined in the organisation's health and safety policy. The policy will set out what the organisation is going to do to manage health and safety and also who is going to do what and how.

Within the policy there should be details of everyone's roles and responsibilities including those with particular functions such as directors, managers, supervisors, safety representatives, workers, fire wardens, first aiders and the competent person. Leaders hold responsibility for making sure that workers and managers are capable of fulfilling their allocated roles. Managers are accountable for ensuring that workers have the necessary training and knowledge to perform their duties and that they understand the information, instruction and training given to them. A key component of this is making sure that all concerned are clear on:

- Their general roles and responsibilities.
- Who specifically is responsible, accountable and competent to undertake particular tasks.

Managers can demonstrate their commitment to health and safety and support for those with responsibilities by using a variety of communication channels to engage the workforce in implementation such as visible behaviour, written material and face-to-face discussions.

To check the extent of implementation of responsibilities and to make necessary adjustments if there is early evidence that requirements are not being met, managers need to monitor performance. If roles and responsibilities for health and safety are included in job descriptions then it can be part of general job appraisal. The overall effectiveness of this is an issue that health and safety practitioners may pursue during health and safety system audits.

STUDY QUESTION

3. Outline activities that can act as drivers to influence ownership of health and safety in an organisation.

(Suggested Answer is at the end.)

Use of Financial Justification

IN THIS SECTION...

The health and safety practitioner needs to be aware of financial considerations, in particular:

- The importance of budgetary responsibilities, including profit, loss and payback analysis.
- The need to identify the responsible budget holder and their influence on health and safety decisions.
- Cost-benefit analysis.
- Sources of information relating to relevant costs.
- The need for budgetary planning (short- and long-term) with regard to health and safety projects.

Significance of Budgetary Responsibility

One of the principal arguments used to justify health and safety initiatives is the financial benefits to be gained. Employers with good health and safety management systems in place are likely to make substantial savings on the cost of accidents that would otherwise have happened. But no business likes to spend money on anything unnecessary, for expenditure means reduced profits or revenues.

Cost-benefit analysis

If the business's income is more than its costs, the business has made a profit; if the business's costs are more than its income, the business has made a loss. The payback period is the amount of time required for the return on an investment to return the sum of the original investment. The business case for a health and safety initiative therefore needs to show that the profit gained from the benefits of the investment will outweigh the loss from the capital expenditure. Justification for health and safety expenditure also needs to substantiate why the cost is necessary, and not avoidable. An obvious reason may be statutory compliance but often a more persuasive business case is required.

The most important part of a cost justification is the perceived benefits resulting from incurring the cost. This is where cost-benefit analysis (see later in this section) can determine the returns or savings expected by making the investment. In addition, incurring the cost may provide indirect benefits that may make the proposal more attractive, although these benefits are often difficult to quantify, e.g. improved morale, better customer perception and image.

Generally, approval becomes easier when more people benefit, and major health and safety initiatives may benefit large sections of the workforce. The benefits of some spending activities, such as Personal Protective Equipment (PPE), can be seen immediately; the benefits of other spending activities, such as training, may take some while to demonstrate a payback. However, wherever appropriate, determine the present value of the expected results, or the future value of the investment, to allow decision-makers to incorporate the time value of money when making the decision.

It might be necessary to consider whether the cost leads to any other follow-up costs. For instance, new ventilation equipment would require additional payments on energy charges and maintenance charges.

Recognising the Responsible Budget Holder

Budget holders are the persons accountable for expenditure from, and income to, particular budgets. They are responsible for the control of their budget and for the general financial administration of their area of responsibility. Consequently, they are the ones responsible for authorising expenditure from their budget.

A budget holder is likely to be a senior manager or director of the business with responsibility for procurement and the associated financial checks and balances. Their perspective is often that their budget is there to be spent as they see fit with the finance function supporting them with budget availability information and making sure that suppliers get paid. The budget holder's priorities are usually geared to the effective operation of their department or business unit and the relevance of health and safety to this may well not be apparent to them.

Health and safety initiatives have to be funded from somewhere and this therefore requires a responsible budget holder to agree to authorise expenditure from their budget. The general arguments used to justify health and safety initiatives may therefore need to be convincingly directed at specific budget holders to influence them to make appropriate decisions.

How this works in practice depends on the organisation's financial arrangements. If the responsibilities of budget holders to allocate resources is specifically defined, then this should include health and safety expenditure. Consequently, health and safety requirements for a particular business unit can easily be matched against the responsible budget holder. If health and safety is less well integrated into business unit management then identifying the budget from which health and safety expenditure will be taken can be more difficult.

The existence of a 'health and safety budget' or 'health and safety contingency fund' can serve to complicate the argument over who the responsible budget holder is. The corporate view may be that budget holders are responsible for funding all health and safety initiatives in their business unit. However, the budget holder may well be of the view that significant health and safety expenditure and initiatives should be centrally funded. These are issues that need to be clarified in the health and safety policy so that it is clear who is responsible for what, and where the responsibility for financial allocation lies.

Cost-Benefit Analysis

There are costs involved with all accidents and losses. There will also be costs involved with accident prevention and risk reduction, in addition to the obvious benefits of such measures. It is possible to spend more on risk treatment than we save by the reduction of the losses. This is why part of risk management is the idea of risk retention.

The cost-benefit graph is illustrated below.

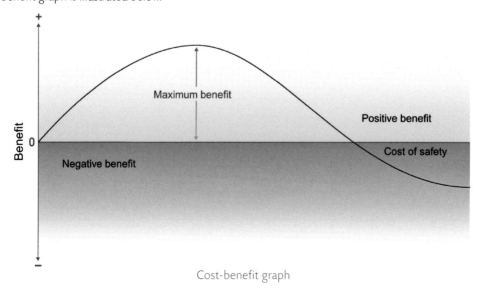

Cost-benefit graph

This graph shows the position where there is maximum benefit.

Cost-benefit analysis is a conceptually simple tool for helping you make a decision as to whether a particular course of action or project is, in fact, viable or cost effective. So, if you are thinking about upgrading risk control measures, you will probably need to justify the request for funding with the aid of a cost-benefit analysis. In its simplest form, it is an entirely economic argument (rather than moral or legal). It is an essential, persuasive tool for the safety practitioner because, not only is it systematic and simple, but it is also commonly used and understood by business people. For this reason, proposed new regulations relating to health and safety are almost always accompanied by a regulatory impact assessment, which contains a cost-benefit analysis to assess financial impact of the proposals on businesses. In the case of regulatory impact assessments, costs may outweigh the benefits for certain industries but, if the proposals become law, you still have to implement their requirements.

In principle, you simply add up all the benefits associated with a programme and then subtract all the costs. In practice, there are a number of complications:

- **Not All Costs and Benefits Can Be Assigned a Reasonably Accurate Financial Value**

  Though we know that intangible things such as 'reputation', 'public/shareholder perception', 'worker morale' and 'worker co-operation and involvement' may have an impact on efficiency, productivity, shareholder investment and sales, their value cannot be fully quantified financially – though it may be possible to propose an estimate.

  ## MORE...

  The HSE's view on cost-benefit analysis, including a checklist, may be found at:

  www.hse.gov.uk/risk/theory/alarpcheck.htm

- **Benefits May Not Be Seen Immediately**

  It may take several years to achieve sufficient benefits to 'break even'. This is known as a payback period. This includes the benefits of reduction in civil liability claims and reductions (or no trend of further rises) in employers' liability insurance premiums.

- **Some Costs and Benefits Are One-Off, Others Are Recurring**

  For example, if your project required the purchase of a new piece of machinery, there is the initial one-off cost of the machine itself, installation, commissioning and any specific training. There are also the annual on-going running costs, such as energy, maintenance, testing, etc. For 'software' projects, such as implementing a safety management system or a behavioural safety programme, you may need to hire extra staff to manage and administer the system, as well as incur costs associated with annual external audits/recertifications.

We have already looked at some typical sources of costs in relation to health and safety accidents. Benefits can be along similar lines (removing a current source of cost is, of course, a future saving, i.e. a benefit of implementing a risk control measure).

There is obviously a cost implication from controlling any kind of risk. **Costs from implementing safety improvement measures** (some of which may have on-going as well as one-off costs) can arise from the following areas:

- Organisational

  These are the costs of any new personnel (salary and training) or perhaps making greater use of an existing resource required to implement and maintain risk control measures. There will also be costs associated with disruption to normal working (temporary staff to cover workers being trained or overtime).

- **Design**

  Reduction of accidents will involve engineering aspects, such as the purchase, fabrication and installation of safety devices, other equipment and any associated software. Safety systems need to be designed and programmes for recording and costing losses will have to be tried out. Costs may also arise from lost production and sales, perhaps due to plant shut-down while equipment is being installed.

- **Planning**

  New safe methods of work, permit-to-work schemes and factory layouts could be considered here.

- **Operational**

  Consideration must be given to the costs of running and maintaining safety systems, maintaining guards, interlocks and software (support, licence renewals), and providing PPE as well as carrying out sampling and testing.

Benefits may arise from issues such as:

- Projected reduction in accidents, with associated savings from less time off and fewer investigations, etc.

- Projected reduction in civil claims.

- Projected reduction in insurance premiums (or reducing the trend of increases due to repeated claims).

- Increased productivity (i.e. reduced cost per unit). This may seem difficult to quantify. However, think about how much time might be saved and translate this to worker-hours. This will give an indication of how much time, and therefore money, may be saved.

You must be prepared to provide and justify estimates of the benefits that you perceive. You will need to analyse your annual accident statistics and consult with your personnel, legal and finance departments to arrive at estimates for some of these benefits.

Initially, you should try to stick to costs and benefits for which you can provide plausible estimates. The more intangible elements for which no financial estimate can be agreed are of more persuasive value. Once you have estimated costs and benefits, you can calculate a projected payback or break-even point. The shorter this is, the better, of course, but some projects are more long term. Even so, do not expect to be greeted with enthusiasm if your projected payback period is much over three years; short payback periods are much more attractive to higher management.

Internal and External Sources of Information

The cost justification for health and safety initiatives is based on the benefits resulting from incurring the cost. The business data and information required for this exercise comes from multiple sources, both external and internal. Costs include:

- New personnel (salary and training) – obtained from internal staffing data and national figures.

- Disruption to normal working (temporary staff to cover workers being trained or overtime) – obtained from past internal staffing costs.

- Purchase, fabrication and installation of safety devices – based on external suppliers' data.

- Lost production and sales – from past balance sheets.

- New safe methods of work and permit-to-work schemes projected from internal and sector figures.

- New factory layouts – based on external suppliers' estimates.

- Running and maintaining safety systems, maintaining guards – projected from internal and suppliers' figures.

- Reduction in accidents, with associated savings – based on internal and national projected accident figures.

- Projected reduction in civil claims – based on past claims experience.

- Projected reduction in insurance premiums – based on internal and national insurance premium trends.

- Increased productivity (i.e. reduced cost per unit) – projected from internal and sector figures.

Short- and Long-Term Budgetary Planning

The business case for health and safety initiatives needs to show that the profit gained from the benefits of the investment will outweigh the loss from the capital expenditure. The cost-benefit analysis determines the returns or savings expected by making the expenditure.

Companies use short-term financial plans to meet budget and investment goals within one fiscal year. These plans can be amended as financial and investment goals change and have a higher degree of certainty compared to long-term plans.

Capital costs for plant and equipment required for health and safety initiatives will appear in the short-term financial plan. Running costs are usually easy to quantify and can be included in the annual budget.

The benefits, or cost savings, from improved health and safety standards tend to be more difficult to quantify and are realised over a longer timescale. Projected savings from reduced accidents and incidents are longer term and difficult to quantify for two main reasons:

- Poor reporting and the general uncertainty in available incident data makes it difficult to estimate potential future losses accurately.

- The full cost of an incident is difficult to quantify and its exact budgetary impact is hard to estimate.

Consequently, the benefits of a health and safety initiative tend to figure in the long-term budgetary plan whereas the loss from capital expenditure will appear very quickly in the short-term budget.

> ## STUDY QUESTION
>
> 4. Identify internal and external sources of information that could be considered when determining costs of health and safety initiatives.
>
> (Suggested Answer is at the end.)

Summary

Role of the Health and Safety Practitioner

We have considered:

- The role of the health and safety practitioner, and discussed the meaning of the term 'competence' in this context.

- The importance of health and safety practitioners evaluating and developing their own practice, and their involvement in mentoring and supporting other employees at all levels.

- The distinction between leadership and management.

- The need to adopt a different management style depending on the situation.

- The influence of the health and safety practitioner on the health and safety management system's development, implementation, maintenance and evaluation.

- Sensible risk management, and the health and safety practitioner's role in enabling work activities as part of it.

- The significance of organisational risk profiling.

- The contribution the health and safety practitioner can make in the achievement of the organisation's objectives.

- The meaning of the term 'ethics' and the practical application of ethical principles within the health and safety profession.

Effective Communication and Negotiation Skills

We have considered:

- The significance of effective communication and the importance of consultation and negotiation when developing a positive health and safety culture.

- The importance of ensuring the ownership of health and safety at all levels within the organisation.

- The need to review health and safety performance on a regular basis and act on feedback received.

- Various methods of communication available to disseminate the health and safety message.

- How to approach the resolution of conflict and the introduction of change.

- The importance of ensuring that all workers understand their roles and responsibilities.

Use of Financial Justification

We have looked at:

- The significance of budgetary responsibility and the importance of recognising who the responsible budget holder is.

- The use of cost-benefit analysis to help in decision-making.

- The varied internal and external sources of information that can be used when assessing cost justification for health and safety initiatives.

- Short- and long-term budgetary planning in the context of health and safety initiatives.

Exam Skills

QUESTION

Explain why organisations often identify the costs of health and safety control measures much more easily than they identify the costs that can arise from poor health and safety standards.   **(10)**

Approaching the Question

Now think about the steps you would take to answer this question:

Step 1:   Read the question carefully.

Step 2:   Next, consider the marks available. This question is worth 10 marks, requires an explanation to answer it, and should take around 15 minutes to answer.

Step 3:   Now, highlight the key words. In this case, the question might look like this:

   Explain why organisations often identify the **costs** of health and safety **control measures** much more easily than they identify the **costs** that can arise from **poor health and safety standards**.   **(10)**

Step 4:   Read the question again to make sure you understand it and have a clear understanding of the two costs it is asking about. (Re-read your notes if you need to.)

Step 5:   The next stage is to develop a plan – there are various ways to do this. A common approach is to consider the information needed to quantify the capital and running costs of providing control measures and the financial losses that would arise from accidents, incidents and occupational ill health. Remember that your answer must be based on the key words you have highlighted.

Step 6:   Now you are in a position to have a go at answering the question. Set out your answer in bullet points with an explanation of each point. Hint – costing potential accidents, incidents and occupational ill health is not an exact science!

When you have finished, have a look at the following comments and guidance.

HINTS AND TIPS

Don't worry too much about the grammar and spelling in your answer, but the examiner MUST be able to understand what you are trying to say. There must be a logical flow to the information you provide and this is where your Answer Plan is so important.

Also remember that the examiner MUST be able to read your handwriting - if they can't read what you have written they can't award you any marks!

Suggested Answer Outline

The examiner would expect you to give an explanation of the relative ease with which the capital and running costs of providing control measures can be quantified compared to the financial losses arising from poor health and safety standards, which are much harder to identify and marks would be available for points similar to the following:

- **Costs of control measures:**
  - Capital costs of controls are known from expenditure.
  - Running costs can be estimated from past operational data.
  - Capital and revenue costs are immediately apparent in the budget.
- **Financial losses from poor health and safety standards:**
  - Savings from reduced accidents and incidents are medium, not short, term.
  - Costs of accidents and ill health are inherently difficult to quantify accurately.
  - Not all loss events are reported.
  - Organisations rarely have effective systems to collect accident costs.
  - To collect accurate cost data on accidents, incidents and occupational ill health requires resources (time and expertise).
  - Financial loss from lost productivity and/or goodwill are even more difficult to estimate.

HINTS AND TIPS

A question that asks you to 'explain' something, expects you to provide an understanding – to make an idea or relationship clear.

Example of How the Question Could Be Answered

Organisations are able to identify the costs of health and safety control measures more easily than the costs arising from poor health and safety standards because control measures are planned and implemented relatively quickly. The capital costs of controls are known from expenditure and the running costs can be estimated from past operational data. Both are relatively easy to obtain and are immediately apparent in the budget.

In comparison, financial losses from poor health and safety standards are likely to occur over a longer period of time and from a number of sources. Savings from reduced accidents and incidents will be medium, not short, term, and can only be estimated from projected figures. The costs of accidents and ill health are inherently difficult to quantify accurately and not all loss events are reported. In addition, organisations rarely have effective systems to collect accident costs because collecting accurate cost data on accidents, incidents and occupational ill health requires additional resources in terms of time and expertise. Finally, the more subtle financial losses that could occur, such as lost productivity and/or goodwill are even more complicated to estimate and quantify to any degree of accuracy.

Reasons for Poor Marks Achieved by Candidates in Exam

An exam candidate would achieve **poor marks** for:

- Not having a clear understanding of the differences between the two types of cost.
- Concentrating on the costs of control measures and not explaining the difficulty of costing poor health and safety standards.
- Failing to realise that some of the costs that arise from poor health and safety standards equate to the savings from reduced accidents and incidents.
- Not including enough information on the difficulty of collecting cost information on accidents, incidents and occupational ill health.
- Failing to include less obvious costs of poor health and safety standards, such as lost productivity and goodwill.

Revision and Examination

The Last Hurdle

Now that you have worked your way through the course material, this section will help you prepare for your NEBOSH examination. This guide contains useful advice on how to approach your revision and the exam itself.

Your NEBOSH Examination

You will need to successfully complete a three-hour examination for each of Units A, B and C, as well as completing Unit DNI, a workplace-based assignment, before you achieve the National Diploma.

Your examination will consist of one exam paper which consists of two parts:

- **Section A** has six short-answer questions worth 10 marks each. These questions are compulsory, and are designed to test your breadth of knowledge across the full range of elements in the syllabus.

- **Section B** has five long-answer questions worth 20 marks each. Only three questions need to be answered from this section which are designed to test your depth of knowledge across the full range of elements in the syllabus.

You are allowed three hours in which to complete the exam and are given ten minutes' reading time before the exam begins.

As a guide, you will need to achieve a minimum of 45% to pass the Unit A, B and C exams, and 50% in the workplace-based assignment (Unit DNI). When you have passed each Unit, you will then be issued with a Unit Certificate, showing a pass grade.

Once you have been awarded a Unit Certificate for all four Units (Units A, B, C and DNI), you will receive an overall grade as follows:

Pass	185 to 239 marks
Credit	240 to 279 marks
Distinction	280 marks or more

The overall mark is calculated by adding together your four Unit Percentage scores.

Remember that your overall grade includes Unit DNI, the workplace-based assignment. Although at this stage of your studies you are quite a way off being ready to attempt the assignment, be aware that you will need to apply what you have learnt throughout your Unit studies when you write your assignment.

Revision Tips

Using the RRC Course Material

You should read through all of your course material once before beginning your revision in earnest. This first read-through should be done slowly and carefully.

Having completed this first revision reading of the course materials, consider briefly reviewing all of it again to check that you understand all of the elements and the important principles that they contain. At this stage, you are not trying to memorise information, but simply checking your understanding of the concepts. Make sure that you resolve any outstanding queries with your tutor.

Remember that understanding the information and being able to remember and recall it are two different things. As you read the course material, you should understand it; in the exam, you have to be able to remember and recall it. To do this successfully, most people have to go back over the material repeatedly.

Re-read the course material and make notes that summarise important information from each element. You could use index cards and create a portable, quick and easy revision aid.

Using the Syllabus Guide

We recommend that you download a copy of the Guide to the NEBOSH National Diploma in Occupational Health and Safety, which contains the syllabus for your course. If a topic is in the syllabus then it is possible that there will be an examination question on that topic.

Map your level of knowledge and recall against the syllabus guide. Look at the **Content** listed for each Unit element in the syllabus guide. Ask yourself the following question:

'If there is a question in the exam about that topic, could I answer it?'

You can even score your current level of knowledge for each topic in each element of the syllabus guide and then use your scores as an indication of your personal strengths and weaknesses. For example, if you scored yourself as 5 out of 5 for a specific topic in Element 1, then obviously you don't have much work to do on that subject as you approach the exam. But, if you scored yourself at 2 out of 5 for a topic in Element 3, then you have identified an area of weakness. Having identified your strengths and weaknesses in this way, you can use this information to decide on the topic areas that you need to concentrate on as you revise for the exam.

Another way of using the syllabus guide is as an active revision aid:

- Pick a topic at random from any of the National Diploma elements.

- Write down as many facts and ideas that you can recall that are relevant to that particular topic.

- Go back to your course material and see what you missed, and fill in the missing areas.

Your revision aim is to achieve a comprehensive understanding of the syllabus. Once you have this, you are in a position to say something on each of the topic areas and attempt any question set on the syllabus content.

Exam Hints

Success in the exam depends on averaging half marks, or more, for each question. Marks are awarded for setting down ideas that are **relevant to the question asked** and demonstrating that you understand what you are talking about. If you have studied your course material thoroughly then this should not be a problem.

One common mistake in answering questions is to go into too much detail on specific topics and fail to deal with the wider issues. If you only cover half the relevant issues, you can only achieve half the available marks. Try to give as wide an answer as you can, without stepping outside the subject matter of the question altogether. Make sure that you cover each issue in appropriate detail in order to demonstrate that you have the relevant knowledge. Giving relevant examples is a good way of doing this.

We mentioned earlier the value of using the syllabus to plan your revision. Another useful way of combining syllabus study with examination practice is to create your own exam questions by adding one of the words you might find at the beginning of an exam question (such as 'explain' or 'identify' or 'outline') in front of the syllabus topic areas. In this way, you can produce a whole range of questions similar to those used in the exam.

Before the Exam

You should:

- Know where the exam is to take place.

- Arrive in good time.

- Bring your examination entry voucher, which includes your candidate number, photographic proof of identity, pens, pencils, ruler, etc. (Remember, these must be in a clear plastic bag or wallet.)

- Bring water to drink and sweets to suck, if you want to.

During the Exam

- Read through the whole exam paper before starting work, if that will help settle your nerves. Start with the question of your choice.

- Manage your time. The exam is three hours long. You should attempt to answer all six questions from Section A and any three questions from Section B in the three hours.

- Check the clock regularly as you write your answers. You should always know exactly where you are, with regard to time.

- As you start each question, read the question carefully. Pay particular attention to the wording of the question to make sure you understand what the examiner is looking for. Note the verbs (command words), such as 'describe', 'explain', 'identify', or 'outline' that are used in the question. These indicate the amount of depth and detail required in your answer. As a general guide:

    - 'Explain' means to provide an understanding. To make an idea or relationship clear.

    - 'Describe' means to give a detailed written account of the distinctive features of a subject. The account should be factual, without any attempt to explain.

    - 'Outline' means to indicate the principal features or different parts of.

    - 'Identify' means to give a reference to an item, which could be its name or title.

- Pay close attention to the number of marks available for each question, or part of a question – this usually indicates how many key pieces of information the examiner expects to see in your answer.

- Give examples wherever possible, based either on your own personal experience, or things you have read about. An example can be used to illustrate an idea and demonstrate that you understand what you are saying.

- If you start to run out of time, write your answers in bullet-point or checklist style, rather than failing to answer a question at all.

- Keep your handwriting neat; if the examiner cannot read what you have written, then they cannot mark it.

- You will not be penalised for poor grammar or spelling, as long as your answers are clear and can be understood. However, you may lose marks if the examiner cannot make sense of the sentence that you have written.

No Peeking!

Once you have worked your way through the study questions in this book, use the suggested answers on the following pages to find out where you went wrong (and what you got right), and as a resource to improve your knowledge and question-answering technique.

Sugg
Ansv

Element A7: The Assessment and Evaluation of Risk

Question 1

Accident and ill-health data may be used to:

- Classify industries according to risk.

- Classify workplaces.

- Classify occupations.

- Consider accident trends.

- Consider parts of the body injured – use of protective clothing.

- Determine hazards in a workplace by using 'cause of injury'.

- Consider where the fault lies.

- Measure the effect of preventive/control measures.

Question 2

Useful internal information sources when assessing risk include:

- Accident and ill-health reports.

- Absence records.

- Maintenance records, which usually show damage incidents.

Question 3

Incidence indicates the number of new cases in a population in relation to the number at risk, whereas **prevalence** indicates the proportion of persons in a given population who have a defined (usually ill-health) condition.

Question 4

The '4 Ps' include:

- **Premises**, including:
    - Access/escape.
    - Housekeeping.
    - Working environment.

- **Plant and substances**, including:
    - Machinery guarding.
    - Local exhaust ventilation.
    - Use/storage/separation of materials/chemicals.

- **Procedures**, including:
    - Permits to work.
    - Use of personal protective equipment.
    - Procedures followed.

- **People**, including:

  - Health surveillance.

  - People's behaviour.

  - Appropriate authorised person.

Question 5

Characteristics of a suitable and sufficient risk assessment:

- It should identify the significant risks arising from, or in, connection with the work.

- The detail in the assessment should be proportionate to the risk.

- Whenever specialist advisers are used, employers should ensure that the advisers have sufficient understanding of the particular work activity they are advising on; this will often require the effective involvement of everyone concerned.

- Risk assessments should consider all those who might be affected by the activities, whether they are workers or others, such as members of the public.

- The risk assessment should indicate the period of time for which it is likely to be valid.

Question 6

(a) Acceptable – no further action required. These risks would be considered by most to be insignificant or trivial and adequately controlled. They are of inherently low risk or can be readily controlled to a low level.

(b) Unacceptable – certain risks that cannot be justified (except in extraordinary circumstances) despite any benefits they might bring. Here we have to distinguish between those activities that we expect those at work to endure, and those we permit individuals to engage in through their own free choice, e.g. certain dangerous sports/pastimes.

(c) Tolerable – risks that fall between the acceptable and unacceptable. Tolerability does not mean acceptable but means that we, as a society, are prepared to endure such risks because of the benefits they give and because further risk reduction is grossly out of proportion in terms of time, cost, etc. In other words, to make any significant risk reduction would require such great cost that it would be out of all proportion to the benefit achieved.

Question 7

(a) Failure is defined as 'the termination of the ability of an item to perform a required function'.

(b) Common mode failure is where two or more components fail in the same way or mode due to a single event or cause, e.g. two or more pairs of braces attached to the same buttons will fail in the same way if the buttons fail. This will not happen if we have one pair of braces and a belt instead. Another example might be a machine where all the components are badly serviced by the same worker with poorly calibrated equipment. Yet another example is where several components are all connected to one other component – if that fails, they all fail in the same way due to that cause.

Question 8

Human reliability can be improved by minimising the number of errors made. This can be achieved by ensuring that the:

- 'Right' person is doing the 'right' job.
- Individual has adequate training and instruction.
- Individual receives appropriate rest breaks.
- Worker–machine interface is ergonomically suitable.
- Working environment is comfortable, e.g. noise, lighting, heating, etc.

Question 9

(a) The following figure shows two components in series:

(b) To calculate the reliability of the series system, the reliabilities are multiplied together:

$R(s) = R(A) \times R(B)$

$= 0.9 \times 0.9$

$= 0.81$

Question 10

The starting point is to logically construct an event tree that begins with pipework failure (f = 0.01/year).

It should look something like this (notice how it is possible to simplify the figure by 'pruning' branches that would not change the outcome):

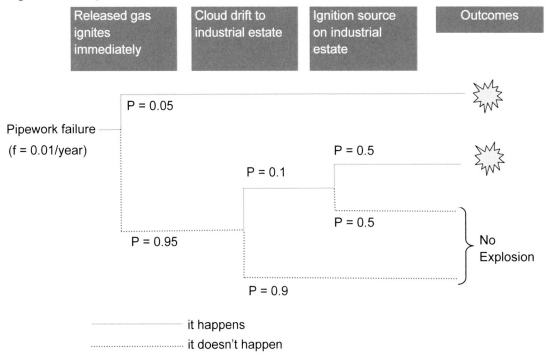

Following the failure of the pipework, there are two possibilities that will result in an explosion. Firstly, the released gas will immediately ignite (probability, P = 0.05) resulting in a jet flame on site. The frequency of this occurrence is calculated simply: 0.01/year × 0.05 = 0.0005/year. Alternatively, this can be expressed as one event in 2,000 years (calculated from 1/0.0005). This gives the answer to the first part of the question. Notice that for this option (ignition on site), the factors of the wind and the industrial estate have no impact; they will not change the result. So, there is no point in further branching of this branch (see figure).

The other option is that the gas does not immediately ignite but drifts off site. If the probability of immediate ignition is 0.05, then the probability of non-ignition and drifting off site must be 0.95 (remember the probability at each node adds up to 1). In most cases, the gas disperses safely but there is a one-in-ten chance (P = 0.1) that it will drift to a nearby industrial estate. Finally, having drifted onto the industrial estate, there is an even chance (P = 0.5) that a source of ignition will cause a vapour cloud explosion or flash fire. Following through this event path on the figure, the frequency of this event occurring is calculated as: 0.01/year × 0.95 × 0.1 × 0.5 = 0.000475/year. This can, alternatively, be expressed as one event every 2,105 years (approximately).

Question 11

Hazard and operability studies are designed for dealing with complex systems, e.g. a large chemical plant. They are carried out by a multidisciplinary team that makes a critical examination of a process to discover any potential hazards and operability problems. A series of guide words are applied to each part of the system to identify the possible consequences of a failure.

Question 12

A **fault tree** identifies the sub-events that are necessary to cause a specified undesired event, such as an accident. Logic gates are used to show how the sub-events combine together to cause the undesired event.

An **event tree** is used to identify the possible outcomes following an undesired event. Both techniques may be used qualitatively and, if suitable data is available, quantitatively.

Element A8: Risk Control

Question 1

The main risk management strategies are:

- Avoidance or elimination.
- Reduction.
- Risk retention – with or without knowledge.
- Risk transfer.
- Risk sharing.

Question 2

Control measures are categorised into three different types:

- Technical – the hazard is controlled or eliminated by designing a new machine or process, or by producing some guarding measure.
- Procedural – such as a safe method of work, e.g. introducing permit-to-work systems as part of a safe system of work.
- Behavioural – involves education and training of operatives, putting up notices and signs, using protective equipment and generally making employees aware of the risks – changing the 'safety culture' of the organisation.

Question 3

The factors to be taken into account when choosing control measures are:

- Long/short term.
- Applicability.
- Practicability.
- Costs.
- Proportionality.
- Effectiveness.
- Legal requirements/standards.
- Competence and training requirements.

Question 4

A permit to work is a formal, written document of authority to undertake a specific procedure and is designed to protect personnel working in hazardous areas or activities.

Question 5

Risk assessment may be used to develop a safe system of work by means of:

- Analysing the task – identifying the hazards and assessing the risks.
- Introducing controls and formulating procedures.
- Instructing and training people in the operation of the system.
- Monitoring and reviewing the system on a regular basis.

Element A9: Organisational Factors

Question 1

(a) **Transformational:**

- People will follow a person who inspires them.
- A person with vision and passion can achieve great things.
- The way to get things done is by injecting enthusiasm and energy.

(b) **Transactional:**

- People are motivated by reward and punishment.
- Social systems work best with a clear chain of command.
- When people have agreed to do a job, a part of the deal is that they pass all authority to their manager.
- The prime purpose of a subordinate is to do what their manager tells them to do.

(c) **Servant:**

- The leader has responsibility for the followers.
- Leaders have a responsibility towards society and those who are disadvantaged.
- People who want to help others best do this by leading them.

Question 2

The benefits of a positive health and safety culture are reflected in indicators of good health and safety performance and include:

- Reduced costs.
- Reduced risks.
- Lower employee absence and turnover rates.
- Fewer accidents.
- Lessened threat of legal action.
- Improved standing among suppliers and partners.
- Better reputation for corporate responsibility among investors, customers and communities.
- Increased productivity, because employees are healthier, happier and better motivated.

Question 3

Effective health and safety leadership will ensure that:

- Instruction, information and training are provided to enable workers to work in a safe and healthy manner.
- Safety representatives are able to carry out their full range of functions.
- The workforce is consulted (either directly or through their representatives) in good time on issues relating to their health and safety and the results of risk assessments.
- Workers are clear who to go to if they have health and safety concerns.
- Line managers regularly discuss how to use new equipment or how to do a job safely.
- Health and safety information is cascaded through the organisation through team meetings, notice boards and other communication channels.

Question 4

The main elements for a risk-based approach for internal control are:

- **Clear Policies and Commitment**

  The board of directors should set a clear policy on risk and internal control. All levels of the company need an understanding of it and should be committed to implementing it.

- **Risk Assessment**

  Identify significant business risks, evaluate and prioritise.

- **Control Environment and Control Activities**

  Develop a clear strategy for dealing with significant risks and define authority, responsibility and accountability.

- **Clear Communication and Reporting Arrangements**

  Establish clear channels of communication, particularly for the periodic reporting of progress regarding business objectives and related risks.

- **Monitoring and Auditing**

  Regularly monitor processes and perform an annual assessment/review of all aspects of internal control processes before making an annual public statement to shareholders on internal control.

Question 5

Internal influences include financial status, production targets, trades unions, and organisational goals and safety culture.

Question 6

External influences include the bodies that are involved in framing legislation and those agencies responsible for its enforcement. Other organisations that may exert an influence on health and safety in the workplace include the courts through their decisions, trade unions by promoting the health and safety of their members, insurance companies by influencing company control measures, professional organisations and various pressure and campaign groups. Public opinion also has a significant influence.

Question 7

The **formal** structure is represented by the company organisation chart, the distribution of legitimate authority, written management rules and procedures, job descriptions, etc. The **informal** structure is represented by individual and group behaviour.

Question 8

The key health and safety issues associated with the following third parties are:

- **Other Employers (Shared Premises)**

  - Co-operate with each other to comply with respective health and safety obligations.

  - Tell other employers about any risks their work activities could present to their employees, both on and off site.

  - Establish who is responsible for what, through communication and co-ordination.

  - Ensure risk assessments consider the risks to others sharing the building or site.

- **Agency Workers**

    - Providers of temporary workers and employers using them should co-operate and communicate with each other.

    - Agreement as to who does what to ensure risks are managed effectively.

    - Before temporary workers start they need to be covered by risk assessments and to know what measures have been taken to protect them.

- **Contractors**

    - Co-operation and co-ordination to make certain that everyone understands the part they need to play to ensure health and safety.

    - Work closely with the contractor to reduce the risks to employees and the contractors themselves.

    - Level of contractor control needed depends on the complexity and the degree of risk associated with the task.

Question 9

The contractor review should include:

- Outcomes and achievements of the contractor.

- Adequacy of procedures in place during the work.

- Consideration of any amendments or additions to the procedures that might be needed.

- Recording the overall performance of the contractor and rating it against established criteria.

- Assembling and providing feedback to the contractor.

Question 10

The functions of a safety representative are to:

- Investigate potential hazards at the workplace.

- Investigate complaints by an employee they represent.

- Make representations to the employer.

- Carry out inspections.

- Consult with HSE inspectors.

- Receive information from inspectors.

- Attend safety committee meetings.

- Inspect the workplace:

    - If they have not inspected it within the previous three months.

    - Where there has been substantial change in the conditions of work.

    - After a notifiable accident, dangerous occurrence or notifiable illness (**RIDDOR**).

Question 11

An employer should consult with employees on:

- Introduction of any measure affecting the health and safety of the employees concerned.
- The appointment of persons nominated to provide health and safety assistance, and assist in emergency procedures (as required by **MHSWR**).
- Any health and safety training or information the employer is required to provide to the employees or the safety representatives.
- The health and safety consequences of the planning and introduction of new technologies in the workplace.
- Provision of any relevant information required on health and safety legislation.

Question 12

Safety circles are small groups of employees – not safety representatives or members of safety committees – who meet informally to discuss safety problems in their immediate working environment. The idea is based on the 'quality circles' concept and allows the sharing of ideas and the suggestion of solutions.

Question 13

A definition should centre on a description of the attitudes, values and beliefs which members of an organisation hold in relation to health and safety, and which, when taken together, produce an organisational culture that can be positive or negative.

Question 14

The most common way to assess safety climate is by using a tool which includes a questionnaire survey asking workers the extent to which they agree or disagree with a number of statements which reflect the management of health and safety.

Question 15

Management commitment can be demonstrated by (three from):

- Managers being seen and involved with the work and correcting health and safety deficiencies.
- Providing resources to carry out jobs safely.
- Ensuring that all personnel are competent.
- Enforcing the company safety rules, and complying with them personally.
- Managers matching their actions to their words.

Question 16

A positive health and safety culture is characterised by:

- Management commitment and leadership.
- High business profile to health and safety.
- Provision of information.
- Involvement and consultation.
- Training.

- Promotion of ownership.

- Setting and meeting targets.

Question 17

The following are needed to effect cultural change:

- Good planning and communication.

- Strong leadership.

- A step-by-step approach.

- Action to promote change.

- Strong worker engagement.

- Ownership at all levels.

- Training and performance measurements.

- Feedback.

Element A10: Human Factors

Question 1

- **Psychology** – a study of the human personality.

- **Sociology** – a study of the history and nature of human society.

Question 2

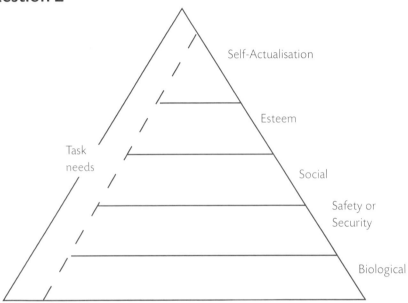

Maslow's hierarchy of needs

Question 3

Human sensory receptors react to danger in the following ways:

- **Sight** – observation of a warning sign.

- **Hearing** – sound of an audible alarm.

- **Taste** – recognition of a toxic substance in food.

- **Smell** – identification of a hazardous gas release.

- **Touch** – identification of a hot surface.

Question 4

Perceptual set: sometimes called a 'mindset'. We have a problem and immediately we perceive not only the problem, but the answer. We then set about solving the problem as we have perceived it. Further evidence may become available, which shows that our original perception was faulty, but we fail to see alternative causes and solutions. This is a basic cause or factor in many accidents and disasters.

Perceptual distortion: the perception of a hazard may be faulty because it gets distorted. Things which are to our advantage always tend to seem more right than those which are to our disadvantage. Management generally tend to have a different perception of hazard from that of workers, and when it affects work rates, physical effort or bonus payments, the worker also suffers from perceptual distortion.

Question 5

The three levels of behaviour in Rasmussen's model are:

- **Skill-based** – the person carries out the operation in automatic mode. Errors occur if there are problems such as machine variation or environmental changes.

- **Rule-based** – the operator is multi-skilled and has a wide selection of well-tried routines which can be used. Errors occur if the wrong alternative is selected or if there is some error in remembering or performing a routine.

- **Knowledge-based** – a person copes with an unknown situation where there are no tried rules or routines. Trial and error is the only method.

Question 6

There were many human errors that contributed to the severity of the incident, including poor hazard analysis, deficiencies in the permit-to-work system, inadequate training in the use of permits and emergency response procedures, and a perceived lack of command by the offshore installation manager on the Tartan rig.

Question 7

Ways that employees could be motivated:

- **Workplace incentive schemes**: encourage employees to work harder in order to receive a payment or benefit.

- **Reward schemes**: offer a reward for improvement or reaching a target in a particular area.

- **Job satisfaction**: some people only require job satisfaction to be motivated. Job satisfaction is very individual to each person.

- **Appraisal schemes**: a formal means of placing value on achievement or effort. It is generally carried out on an annual basis and the results may be used to determine the level of a pay rise or a promotion.

Question 8

Formal groups are established to achieve set goals, aims and objectives. They have clearly defined rules, structures and channels of communication.

Informal groups superimpose on the organisation an informal structure of communication links and functional working groups. These cross all the barriers of management status and can be based on family relationships, out-of-work activities, experience or expertise.

Question 9

Shift work can be very demanding on an individual and can affect their performance in the following ways:

- **Fatigue and stress** – poorer performance on tasks requiring attention, decision-making or high levels of skill.

- **Sleep loss and sleep debt** – lower levels of alertness, and reduced levels of productivity and attention.

- **Health problems** – asthma, allergic reactions and respiratory problems tend to be worse at night, and so it is likely that performance will be affected where an individual's health is affected.

- **Social life/family life** – work performance may be affected if the individual is unhappy at home due to the constraints of shift work.

- **Natural circadian rhythm** – when working nights, the body still reduces body temperature in the early hours of the morning, reduces blood pressure and stops digestion which can lead to an individual feeling sleepy and less alert.

Question 10

There are various methods of payment for work and terms of employment, and these may all affect performance in different ways:

- **Piecemeal workers** – need to work quickly because they are paid by the amount of work they do; speed is important, but safety is also an issue, because if they injure themselves they won't be able to work and will not get paid.

- **Permanent contract employees** – get paid whether they are at work or sick (in general), so safety or performance may not be an incentive for them. There may, however, be incentive schemes in place which reward good performance, safe working and/or good attendance, and these may affect an individual's performance.

- **Short-term contract workers** – generally need to perform well in order for their contract to be renewed (unless there is a shortage of workers). This means that there is pressure on the individual to perform to their best ability.

Question 11

Ergonomics is the study of the relationship between workers and their environment, ensuring a good 'fit' between people and the things they use. Essentially, it involves 'fitting the task to the worker' rather than 'fitting the worker to the task'. The order of operations and work practices can be modified so that each person is working to full efficiency. Poorly designed work equipment and unsafe practices may result in injury and occupational ill health. These may include:

- Equipment not suited to body size.

- Operator not able to readily see and hear all that they need to.

- Lack of understanding of the information that is presented to the employee.

- Equipment or system causing discomfort if used for any length of time.

Question 12

The following features are present in an ergonomically designed crane cab control system:

- The controls are within easy reach of the driver and are moved in a straight line to allow ease and delicacy of control.

- The seat is adjustable so that the driver has a good view of the operations.

- The environment of the cab protects the driver from dust and fumes, etc.

Question 13

The steps of a behavioural change programme relating to safety are:

Step 1: Identify the specific observable behaviour that needs changing, e.g. to increase the wearing of hearing protectors in a high-noise environment.

Step 2: Measure the level of the desired behaviour by observation.

Step 3: Identify the cues (or triggers) that cause the behaviour and the consequences (or pay offs) (good and bad) that may result from the behaviour.

Step 4: Train workers to observe and record the safety critical behaviour.

Step 5: Praise/reward safe behaviour and challenge unsafe behaviour.

Step 6: Feed back safe/unsafe behaviour levels regularly to workforce.

Element A11: The Role of the Health and Safety Practitioner

Question 1

The concept of sensible risk management aims to balance the growing 'risk averse' attitude of society toward innovation and development and involves:

- Ensuring that workers and the public are properly protected.

- Enabling innovation and learning, not stifling them.

- Ensuring that those who create risks manage them responsibly and understand that failure to do so is likely to lead to robust action.

- Providing overall benefit to society by balancing benefits and risks, with a focus on reducing significant risks:

    - Those with serious consequences.

    - Those which arise more often.

- Enabling individuals to understand that as well as the right to protection, they also have to exercise responsibility.

It is **not** about:

- Reducing protection of people from risks that cause real harm.

- Scaring people by exaggerating or publicising trivial risks.

- Stopping important recreational and learning activities for individuals where the risks are managed.

- Creating a totally risk-free society.

- Generating useless paperwork mountains.

Question 2

Examples of how a safety practitioner would be expected to adhere to ethical principles are (any five from the following):

- Abide by relevant legal requirements.

- Give honest opinions.

- Maintain their competence.

- Undertake only those tasks they believe themselves to be competent to deal with.

- Accept professional responsibility for their work and make those who ignore their professional advice aware of the consequences.

- Not bring the professional body into disrepute, injure the professional reputation or business of others or behave in a way that may be considered inappropriate.

- Not use their membership or position within the organisation or Institution improperly for commercial or personal gain.

Question 3

Activities that can act as drivers to influence ownership of health and safety in an organisation:

- **Participation** by employees supports risk control by encouraging their 'ownership' of health and safety policies. It establishes an understanding that the organisation as a whole, and people working in it, benefit from good health and safety performance. Pooling knowledge and experience through participation, commitment and involvement means that health and safety really becomes 'everybody's business'.

- **Management accountability** means that managers are judged on how well or effectively they carry out the duties they are responsible for. Consequently, their credibility relies on taking on board their responsibilities and accounting for or explaining the actions taken (or not taken) to senior managers.

- **Consultation** can occur at each stage of the health and safety management system:

    - Consult workers or their representatives during the planning and organisation of training.

    - Involve and consult workers and their representatives during the implementation process, by ensuring there are systems in place that allow workers to raise concerns and make suggestions.

    - Involve the workforce in setting and monitoring performance measures and encourage them to monitor their own work area.

    - Discuss plans for review with workers or their representatives, use information from safety representatives' inspections to feed into the review and discuss the review findings with workers or their representatives.

- **Feedback** on success and failure is an essential element in motivating employees to maintain and improve performance. Successful organisations emphasise positive reinforcement and concentrate on encouraging progress on those indicators which demonstrate improvements in risk control. This also encourages identification with and ownership of the health and safety programme.

Question 4

Sources of information to consider include:

- New personnel (salary and training) – obtained from internal staffing data and national figures.

- Disruption to normal working (temporary staff to cover workers being trained or overtime) – obtained from past internal staffing costs.

- Purchase, fabrication and installation of safety devices – based on external suppliers' data.

- Lost production and sales – from past balance sheets.

- New safe methods of work and permit-to-work schemes – projected from internal and sector figures.

- New factory layouts – based on external suppliers' estimates.

- Running and maintaining safety systems, maintaining guards – projected from internal and suppliers' figures.

- Reduction in accidents, with associated savings – based on internal and national projected accident figures.

- Projected reduction in civil claims – based on past claims experience.

- Projected reduction in insurance premiums – based on internal and national insurance premium trends.

- Increased productivity (i.e. reduced cost per unit) – projected from internal and sector figures.